INTERNET THEORY, TECHNOLOGY AND APPLICATIONS

U.S. TECHNOLOGICAL ENDEAVORS

EXAMINATIONS OF DIGITAL TRADE AND SEMICONDUCTOR MANUFACTURING

INTERNET THEORY, TECHNOLOGY AND APPLICATIONS

Additional books in this series can be found on Nova's website under the Series tab.

Additional e-books in this series can be found on Nova's website under the eBooks tab.

INTERNET THEORY, TECHNOLOGY AND APPLICATIONS

U.S. TECHNOLOGICAL ENDEAVORS

EXAMINATIONS OF DIGITAL TRADE AND SEMICONDUCTOR MANUFACTURING

BEVERLY HOWARD
EDITOR

New York

Copyright © 2017 by Nova Science Publishers, Inc.

All rights reserved. No part of this book may be reproduced, stored in a retrieval system or transmitted in any form or by any means: electronic, electrostatic, magnetic, tape, mechanical photocopying, recording or otherwise without the written permission of the Publisher.

We have partnered with Copyright Clearance Center to make it easy for you to obtain permissions to reuse content from this publication. Simply navigate to this publication's page on Nova's website and locate the "Get Permission" button below the title description. This button is linked directly to the title's permission page on copyright.com. Alternatively, you can visit copyright.com and search by title, ISBN, or ISSN.

For further questions about using the service on copyright.com, please contact:
Copyright Clearance Center
Phone: +1-(978) 750-8400   Fax: +1-(978) 750-4470   E-mail: info@copyright.com.

NOTICE TO THE READER

The Publisher has taken reasonable care in the preparation of this book, but makes no expressed or implied warranty of any kind and assumes no responsibility for any errors or omissions. No liability is assumed for incidental or consequential damages in connection with or arising out of information contained in this book. The Publisher shall not be liable for any special, consequential, or exemplary damages resulting, in whole or in part, from the readers' use of, or reliance upon, this material. Any parts of this book based on government reports are so indicated and copyright is claimed for those parts to the extent applicable to compilations of such works.

Independent verification should be sought for any data, advice or recommendations contained in this book. In addition, no responsibility is assumed by the publisher for any injury and/or damage to persons or property arising from any methods, products, instructions, ideas or otherwise contained in this publication.

This publication is designed to provide accurate and authoritative information with regard to the subject matter covered herein. It is sold with the clear understanding that the Publisher is not engaged in rendering legal or any other professional services. If legal or any other expert assistance is required, the services of a competent person should be sought. FROM A DECLARATION OF PARTICIPANTS JOINTLY ADOPTED BY A COMMITTEE OF THE AMERICAN BAR ASSOCIATION AND A COMMITTEE OF PUBLISHERS.

Additional color graphics may be available in the e-book version of this book.

Library of Congress Cataloging-in-Publication Data

ISBN: 978-1-53610-547-6

Published by Nova Science Publishers, Inc. † New York

CONTENTS

Preface		vii
Chapter 1	Digital Trade and U.S. Trade Policy *Rachel F. Fefer, Shayerah Ilias Akhtar and Wayne M. Morrison*	1
Chapter 2	Digital Economy and Cross-Border Trade: The Value of Digitally-Deliverable Services *Jessica R. Nicholson and Ryan Noonan*	53
Chapter 3	U.S. Semiconductor Manufacturing: Industry Trends, Global Competition, Federal Policy *Michaela D. Platzer and John F. Sargent Jr.*	85
Index		117

PREFACE

As the rules of global Internet develop and evolve, digital trade has risen in prominence on the global trade and economic agenda, but multilateral trade agreements have not kept pace with the complexities of the digital economy. The economic impact of the Internet is estimated to be $4.2 trillion in 2016, making it the equivalent of the fifth-largest national economy. According to one source, the volume of global data flows grew 45-fold from 2005 to 2014, faster than international trade or financial flows. Congress has an important role to play in shaping global digital trade policy, from oversight of agencies charged with regulating cross-border data flows to shaping and considering legislation to implement new trade rules and disciplines through ongoing trade negotiations, and also working with the executive branch to identify the right balance between digital trade and other policy objectives, including privacy and national security. This book discusses the role of digital trade in the U.S. economy, barriers to digital trade, digital trade agreement provisions, and other selected policy issues. It also discusses the digital economy and cross-border trade, and U.S. semiconductor manufacturing.

In: U.S. Technological Endeavors
Editor: Beverly Howard

ISBN: 978-1-53610-547-6
© 2017 Nova Science Publishers, Inc.

Chapter 1

DIGITAL TRADE AND U.S. TRADE POLICY*

Rachel F. Fefer, Shayerah Ilias Akhtar and Wayne M. Morrison

SUMMARY

As the rules of global Internet develop and evolve, digital trade has risen in prominence on the global trade and economic agenda, but multilateral trade agreements have not kept pace with the complexities of the digital economy. The economic impact of the Internet is estimated to be $4.2 trillion in 2016, making it the equivalent of the fifth-largest national economy. According to one source, the volume of global data flows grew 45-fold from 2005 to 2014, faster than international trade or financial flows. Congress has an important role to play in shaping global digital trade policy, from oversight of agencies charged with regulating cross-border data flows to shaping and considering legislation to implement new trade rules and disciplines through ongoing trade negotiations, and also working with the executive branch to identify the right balance between digital trade and other policy objectives, including privacy and national security.

Digital trade includes end-products like movies and video games and services such as e-mail. Digital trade also enhances the productivity and overall competitiveness of an economy. According to the U.S. International Trade Commission, U.S. domestic and international digital

---

* This is an edited, reformatted and augmented version of a Congressional Research Service publication, R44565, dated July 15, 2016.

trade added 3.4 - 4.8% ($517.1-$710.7 billion) to the U.S. gross domestic product (GDP) in 2011. The Department of Commerce found that in 2014, digitally-delivered services accounted for more than half of U.S. services trade.

The increase in digital trade also raises new challenges in U.S. trade policy, including how to best address new and emerging trade barriers. As with traditional trade barriers, digital trade constraints can be classified as tariff or nontariff barriers. In addition to high tariffs, barriers to digital trade may include localization requirements, cross border data flow limitations, intellectual property rights (IPR) infringement, unique standards or burdensome testing, filtering or blocking, and cybercrime exposure or state-directed theft of trade secrets.

Digital trade issues often overlap and cut across policy areas, including IPR and national security; this raises questions for Congress as it weighs different policy objectives. The Organization for Economic Cooperation and Development (OECD) points out three potentially conflicting policy goals in the Internet economy: (1) enabling the Internet; (2) boosting or preserving competition within and outside the Internet; and (3) protecting privacy and consumers more generally.

While no comprehensive agreement on digital trade exists in the World Trade Organization (WTO), other WTO agreements do cover some aspects of digital trade. Recent bilateral and plurilateral agreements have begun to address digital trade rules and barriers more explicitly. For example, the potential Trans-Pacific Partnership (TPP), Transatlantic Trade and Investment Partnership (T-TIP), and plurilateral Trade in Services Agreement (TiSA) are expected to address digital trade to varying degrees. Digital trade norms are also being discussed in forums such as the Group of 20 (G-20), the OECD, and the Asia-Pacific Economic Cooperation (APEC), providing the United States with multiple opportunities to engage in and shape global developments.

Congress has an interest in ensuring the global rules and norms of the Internet economy are in line with U.S. laws and norms, and in establishing a U.S. trade policy on digital trade that advances U.S. interests.

INTRODUCTION

The Internet-driven digital revolution is causing fundamental change to the global economy leading not only to new modes of communication and information-sharing, new business models, and new sources of job growth, but also to new policy questions and concerns. According to a report by the McKinsey Global Institute, globalization has entered "a new era defined by data flows that transmit information, ideas, and innovation."[1] Another report noted "information is currency... Information is also the building block of the digital economy."[2] As digital information increases in importance in the U.S. economy, issues related to digital trade have become of growing interest in trade negotiations.

The U.S. International Trade Commission (ITC) broadly defines digital trade as "U.S. domestic commerce and international trade in which the Internet and Internet-based technologies play a particularly significant role in ordering, producing, or delivering products and services."[3] Thus, digital trade not only includes end-products like movies and video games, but also provides the means to enhance the productivity and overall competitiveness of an economy. Examples of digital trade include: orders placed on an e-commerce website; information streams needed by manufacturers to manage global value chains; communication channels such as email and voice over Internet protocol (VoIP); and financial data and transactions relied on for online purchases or electronic banking.

The rules governing digital trade are evolving as governments across the globe experiment with different approaches and try to balance diverse policy priorities and objectives. Barriers to digital trade, such as infringement of intellectual property rights (IPR), national security measures, or industrial policies, often overlap and cut across sectors. Digital trade issues have been in the spotlight recently, due in part to heightened concerns over data privacy and an increasing number of cybertheft incidents that have affected U.S. consumers and companies. These concerns may affect the general U.S. interest in promoting cross-border data flows. Congress has an interest in ensuring the global rules and norms of the Internet economy are in line with U.S. laws and norms.

Trade negotiators continue to explore ways to address digital issues in trade agreements, including the proposed Trans-Pacific Partnership (TPP), which contains the most advanced disciplines to date on digital trade barriers. Congress has an important role in shaping digital trade policy, from oversight of agencies charged with regulating cross-border data flows and of ongoing trade negotiations, to working with the executive branch to identify the right balance between digital trade and other policy objectives, including privacy and national security concerns.

This report discusses the role of digital trade in the U.S. economy, barriers to digital trade, digital trade agreement provisions, and other selected policy issues.

ROLE OF DIGITAL TRADE IN THE U.S. AND GLOBAL ECONOMY

The Internet not only has become a facilitator of existing international trade in goods and services, but is itself a platform for new digitally-originated services. The Internet is enabling technological shifts that are transforming businesses. According to a study by the Boston Consulting Group, the global economic impact of the Internet is estimated to be $4.2 trillion in 2016, and would rank as the fifth-largest national economy in the world. Some estimates indicate that gross domestic product (GDP) in developed countries is 5% to 9% higher annually (largely through increased productivity and lower costs) than it would be without the Internet, while in developing countries the Internet has an even larger impact, adding 15% to 25% to GDP per year.[4] According to one estimate, the volume of global data flows is growing faster than trade or financial flows, as *Figure 1* illustrates, growing 45-fold from 2005 to 2014.

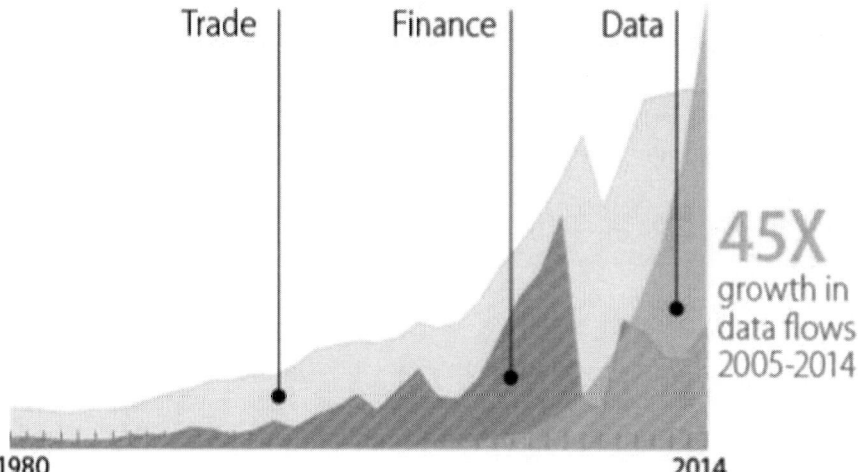

Source: McKinsey Global Institute, Digital Globalization: The New Era of Global Flows, March 2016.

Figure 1. Growth in Global Data Flows.

The increase in digital trade parallels the growth in Internet usage globally. Today, there are more than 2.7 billion Internet users worldwide.[5] World Bank estimates are even higher, showing that Internet users tripled since 2005 to 3.2 billion in 2015, representing 60% of people globally.[6] The

Organization for Economic Cooperation and Development (OECD) reports that in 2014, on average 95% of enterprises in OECD countries had a broadband connection and 76% had a website or homepage.[7] In the United States, 92% of the population uses the Internet, according to one estimate.[8] While 75% of U.S. households use wired Internet access, an increasing number (20%) are relying on mobile Internet access, with low-income households more likely to rely on wireless (29%). While the percentage of American consumers relying on a desktop or laptop at home is declining (34% and 46%, respectively), they increasingly are turning to an array of devices from smartphones to wearable devices for Internet access, according to one estimate.9 Each day, companies and individuals depend on the Internet to communicate and transmit data via various media and channels that continue to expand (see *Figure 2*).

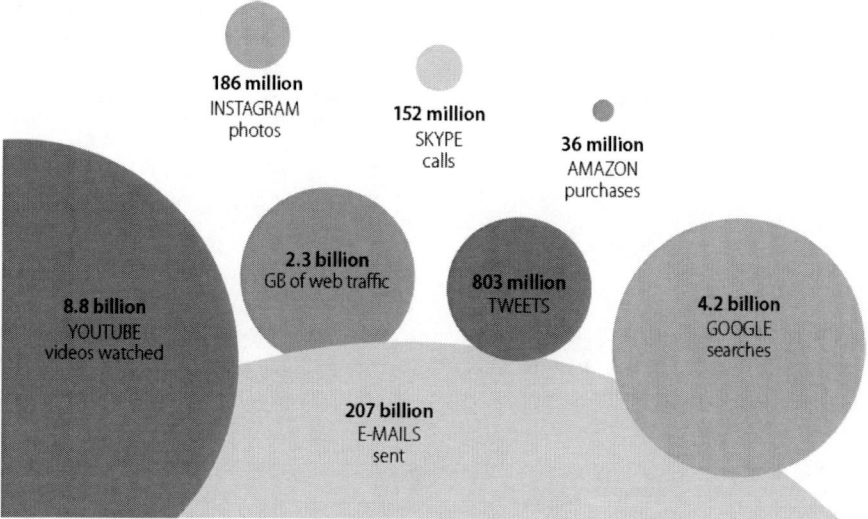

Source: The World Bank Group, World Development Report 2016: Digital Dividends, 2016, p. 6, http://www.worldbank.org/en/publication/wdr2016.

Figure 2. A Typical Day in the Life of the Internet.

According to one study, global cross-border Internet traffic grew 60% a year between 2002 and 2012.[10] Some analysts also conclude that most of the bilateral trade in data-intensive sectors takes place between countries in the OECD, and find a correlation with foreign direct investment (FDI).[11] OECD countries are also more likely to have the necessary underlying infrastructure to support high data flows.

Cross-border data and communication flows are themselves part of digital trade; they also facilitate trade and the flows of goods, services, people, and finance, which together are the drivers of globalization and interconnectedness. According to one estimate, worldwide data and communication flows have grown more than sevenfold from 2008 to 2013 (See *Figure 3*).[12] The highest levels reportedly are those flows between the United States and Western Europe, Latin America, and China.

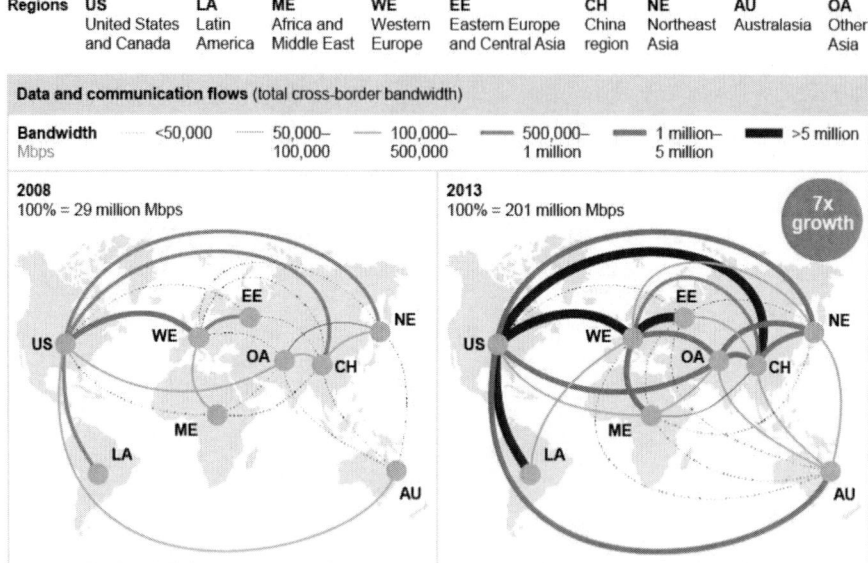

Source: McKinsey Global Institute, Global Flows in a digital age: How trade, finance, people, and data connect, April 2014, p. 13, http://www.mckinsey.com/business-functions/strategy-and-corporate-finance/our-insights/globalflows-in-a-digital-age.

Notes: Circle indicates size of increase.

Figure 3. Data and Communication Flows between Regions.

Powering all these connections and data flows are underlying information and communication technologies (ICT).[13] The Business Software Alliance (BSA) estimates that over $2 trillion are spent each year on information technologies and services. ICT spending is a large and growing component of the international economy. Globally, ICT spending is growing at a compounded annual rate of 3.4%, and is forecasted to be more than $4 trillion in 2017. In 2012, the United States was estimated to be the largest purchaser of ICT at $942 billion.[14]

ICT services are outpacing the growth of international trade in ICT goods. According to the OECD, ICT services increased fourfold between 2001 and 2013. The United States is the fourth largest OECD exporter of ICT services, after Ireland, India, and Germany.[15]

Digitization of Trade Flows

As the Internet and technology continue to develop, increasing digitization affects finance and data flows, as well as the movement of people and goods. Beyond simple communication, McKinsey describes three major ways digital technologies affect global trade flows:[16]

1. Digitization creates new digital goods and services. Digital technology enables innovation. By transforming, and often replacing, traditional goods and services, or the need for people to travel, new products are conceived (e.g., e-books, remote or virtual office for collaboration, tele-medicine, online education or banking).
2. Digitization enhances physical flows through "digital wrappers." Digital wrappers add value by raising productivity, and/or lowering the costs and barriers related to flows of traditional goods and services (e.g., radio-frequency identification (RFID) tags for supply chain tracking, data files used in 3-D printing (or additive manufacturing), cars automatically transmitting data, the "Internet of Things" to connecting devices or objects).[17]
3. Digitization provides platforms that serve as intermediaries for production, exchange, and consumption.[18] Intermediary platforms include not only those used in e-commerce, but also for social media, crowd funding, cloud computing, search engines, big data analytics, sharing services (e.g., car or accommodation sharing such as Uber or Airbnb), and mobile "apps," or applications.[19]

Economic Impact of Digital Trade

The World Bank identifies three buckets of "digital dividends," benefits that result from using digital technologies: (1) inclusion through increased access and reach; (2) efficiency through automation and coordination; and (3) innovation driving new businesses.[20] These dividends can accrue to businesses, individuals, and governments. Firms that use the Internet more intensively show higher productivity and tend to be larger, faster-growing, and more skill and export intensive. An increase in Internet usage is also associated with an increase in the number and value of products being traded. Drilling down further into the economic benefits of digital trade, the ITC identified specific benefits for consumers and workers (e.g., reduced prices, increased selection, and higher employment) and businesses (e.g., increased efficiency, productivity, output, exports, and sales).[21]

According to some estimates, digital trade, including both U.S. domestic commerce and international trade, increased U.S. GDP by an estimated 3.4% - 4.8% ($517.1-$710.7 billion) in 2011. In addition, U.S. real wages increased by an estimated 4.5% - 5.0% and total U.S. employment was higher by 2.4 million full-time equivalents (FTEs) as a result of digital trade.[22]

Looking at digital trade in an international context, global cross-border e-commerce from online sales (excluding domestic sales) was estimated to be 10% to 15% of total e-commerce in 2014.[23] In the same year, the United States exported $399.7 billion in digitally-deliverable services, and imported $240.8 billion, creating a surplus of $158.9 billion. Digitally-delivered services accounted for more than half of all U.S. services trade, according to the Department of Commerce.[24] Other estimates show that, without the Internet, the costs of U.S. imports and exports would have been an average of 26 times higher.[25] Furthermore, these estimates do not quantify the additional benefits of digitization upon business efficiency and productivity, or of increased customer and market access, which enable greater volumes of international trade.

Digital platforms can minimize costs and enable small and medium-sized enterprises (SMEs) to grow through extended reach or integrating into a global value chain (GVC) (see *text box*). As a result, more firms are able to conduct business in global markets (or are more willing to do so), while the digitization of customs and border control mechanisms helps simplify and speed delivery of goods to customers. A study of U.S. SMEs on the e-commerce platform eBay found that 97% export while that number is a full 100% in countries as diverse as Peru and Ukraine.[26]

Another study of SMEs estimated that the Internet is a net creator of jobs, with 2.6 jobs created for every job that may be displaced by Internet technologies; companies that use the Internet intensively effectively doubled the average number of jobs.[27] However, the costs of digital trade can be concentrated on particular sectors (see next section).

> **Idaho Company Thrives with Digital Trade**
>
> TSheets co-founders Matt Rissell and Brandon Zehm created an Internet cloud-based, employee-time-tracking solution that worked with QuickBooks. Started in 2006, the company has since hired 60 employees, expanded into 63 countries, and was named Idaho's Innovative Company of the Year by the Idaho Technology Council. The company uses Google services for online advertising and customer engagement, analytics, document storage, and even to enhance their own products. "Because of the Internet and the tools available to us, we've been able to grow an international company based in Boise, Idaho," Matt says.[28]

Digitization Challenges

The U.S. economy may only be realizing 18% of its digital potential and it is doing so unevenly across sectors and populations.[29] Industries, such as media and those in urban centers, account for a larger share of the benefits. Many in business and research communities are only beginning to understand how to take advantage of the vast amounts of data being collected every day. Some experts estimate digitization could add another $2.2 trillion a year to the U.S. GDP by 2025.[30]

Additionally, sources of "e-friction" or obstacles can prevent consumers, companies, and countries from realizing the full benefits of the online economy.[31] Causes of e-friction can fall into four categories: infrastructure; industry; individual; and information. Government policy can influence e-friction, from investment in infrastructure and education to regulation and online content filtering. According to some experts, economies with lower amounts of e-friction may be associated with larger digital economies.[32]

While there are numerous positive digital dividends, there are also potential negative and uneven results across populations, such as the displacement of unskilled workers, an imbalance between companies with and without Internet access, and a tendency for some to use the Internet to establish monopolies.[33] While new technologies and new business models present opportunities to enhance efficiency and expand revenues, innovate faster, and achieve other benefits, new challenges also arise with the disruption of supply chains, labor markets, and some industries.

The World Bank identified policy areas to ensure, and maintain, the potential benefits of digitization. Policy areas include: establishing a favorable and competitive business climate, developing strong human capital, ensuring good governance, investing to improve both physical and digital infrastructure, and raising digital literacy skills. According to the World Economic Forum Competitiveness Rankings which looks at technological adoption and ICT use, the United States is ranked seventeenth.[34] With the rapid pace of technology innovation, more jobs may become automated, with digital skills becoming a foundation for economic growth, for individual workers, companies, and national GDP.[35]

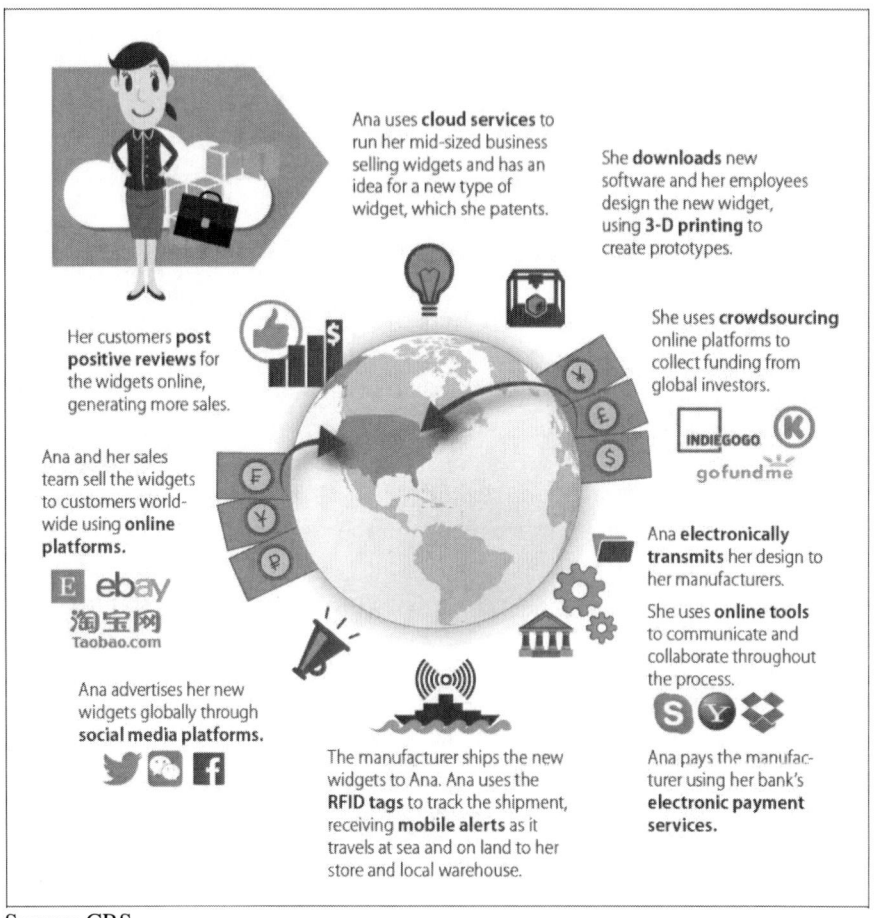

Source: CRS.

Notes: The above graphic is illustrative only and is not based on a real business or reflective of all aspects of digital trade.

Figure 4. What is Digital Trade? (Examples of international digital trade)

DIGITAL TRADE BARRIERS AND POLICY ISSUES

Policies that affect digitization in any one country's economy can have consequences beyond its borders, and because the Internet is a global "network of networks," the state of a country's digital economy can have global ramifications. Protectionist policies may erect barriers to digital trade, or damage trust in the underlying digital economy, and can result in the fracturing, or so called balkanization, of the Internet, lessening any gains. What some policymakers see as protectionist, however, others may view as necessary to protect domestic interests. Ensuring a free and open Internet is a stated policy priority for the U.S. government.[36] Like other crosscutting policy areas, such as cybersecurity, no one federal entity has policy primacy on all aspects of digital trade and the United States has taken a sectoral approach to regulating digitization.

Ensuring a Free and Open Internet[37]

Ensuring a free and open Internet is a policy priority according to the U.S. Department of State. The U.S. Trade Representative aims to promote this position through global trade.

"...Digital freedom must triumph over digital protectionism. Around the world, policies restricting the free flow of data and the openness of the Internet are on the rise, threatening to effectively balkanize the Internet... Policies requiring companies to store data locally present another serious threat, making costs prohibitively high for many small businesses, curtailing access to global services, and stifling innovation...

Above and beyond its impact on commerce, digital freedom goes to the heart of what it means to live in the information age. Ensuring that the rules of the road for global trade promote the free flow of information and resist artificial barriers has broad ramifications. When data flows are obstructed, everyone from the immigrant keeping in touch with relatives, to the work-from-home entrepreneur connecting with customers, to the aspiring high school blogger can be affected." – U.S. Trade Representative, Ambassador Michael Froman.

The Department of Commerce National Telecommunications and Information Administration notes key U.S. policies that enable a strong digital economy in this country include: (1) connecting and empowering users; (2) trusting the private sector and protecting online platforms; (3) a strong and balanced approach to intellectual property that fosters innovation while recognizing "fair use"; and (4) a multi-stakeholder consensus-based process for Internet governance.[38] The absence of similar policies, or the existence of opposing ones, outside the United States can lead to trade barriers that hinder or block the flow of digital trade.

The Department of Commerce launched a Digital Economy Agenda that identifies four pillars:[39]

1. Promoting a free and open Internet worldwide, because the Internet functions best for our businesses and workers when data and services can flow unimpeded across borders;
2. Promoting trust online, because security and privacy are essential if electronic commerce is to flourish;
3. Ensuring access for workers, families, and companies, because fast broadband networks are essential to economic success in the 21st century; and
4. Promoting innovation, through smart intellectual property rules and by advancing the next generation of exciting new technologies.

The Commerce Secretary launched specific efforts to support the Digital Economy Agenda including a Digital Economy Board of Advisors from across sectors and a pilot digital attaché program under the foreign commercial service to help U.S. businesses navigate regulatory issues and overcome trade barriers to e-commerce exports.[40]

As with traditional trade barriers, digital trade constraints can be classified as tariff or nontariff barriers. Tariff barriers may be imposed on imported goods used to create ICT infrastructure that make digital trade possible or on the products that allow users to connect, while nontariff barriers, such as discriminatory regulations or local content rules, can block or limit different aspects of digital trade. Often, such barriers are intended to protect domestic producers and suppliers. The ITC estimated that removing foreign barriers to digital trade could increase annual U.S. real GDP by 0.1–0.3% ($16.7–$41.4 billion), increase U.S. wages up to 1.4%, and add up to 400,000 U.S. jobs in certain digitally intensive industries.[41]

2015 U.S. Digital Trade Negotiating Objectives

Congress enhanced its digital trade objectives for U.S. trade negotiations in the Bipartisan Congressional Trade Priorities and Accountability Act of 2015 (P.L. 114-26), or Trade Promotion Authority (TPA), signed into law in June 2015.[42] Congress recognized the importance of digital trade and removing related barriers when it passed TPA. TPA 2015 objectives related to digital trade direct the Administration to negotiate agreements that:

- Ensure application of existing WTO commitments to digital trade environment, ensuring no less favorable
- treatment to physical trade;
- Prohibit forced localization requirements and restrictions to digital trade and data flows;
- Keep electronic transmissions duty-free; and
- Ensure relevant legitimate regulations are as least trade restrictive as possible.

Tariff Barriers

Tariffs may impede goods trade at the border by raising the prices of U.S. products as costs are passed to end customers, thus limiting market access for U.S. exporters. Quotas may limit the number or value of foreign goods, persons, suppliers, or investments allowed in a market.

Global exports of ICT goods reached $1.6 trillion in 2013, and production is increasingly concentrated in a few countries, with China (32% of ICT good exports), United States (9%), and Singapore (8%) ranking at the top.[43] For example, semiconductors, a key component in many electronic devices, are a top U.S. ICT export. They were the number three U.S. manufactured export over the last five years with 2014 sales of $172.9 billion.[44] U.S. ICT services are often inputs to final demand products that may be exported by other countries, such as China. While the United States is a major exporter and importer of ICT goods, tariffs are not levied on many of the products due to free trade agreements (FTAs) and the World Trade Organization Information Technology Agreement (see below). Tariffs may still serve as trade barriers for those countries or products not covered by existing FTAs or the WTO ITA.

ICT Goods Tariff Barriers: Selected Examples

Brazil, Mexico, and Vietnam are key participants in the ICT goods market and impose high tariffs on non-FTA partners. According to the United Nations Statistics Division, in 2015 Brazil reported $1.3 billion in medical ICT equipment imports such as electrocardiographs, ultrasound devices, and magnetic resonance imaging devices,[45] despite tariffs of up to 16% on these products.[46]

In 2014, Vietnam reportedly imported $10.3 billion worth of electronic integrated circuits (microchips) and associated parts, including approximately 4% or $398 million from the United States.[47] While Vietnam imposes no tariffs on these product categories, several ICT items in Vietnam's tariff schedule have high applied rates, including multiple categories of radio equipment, which have an applied rate as high as 30% according to the WTO.[48]

Mexico and Vietnam are both members of the proposed Trans-Pacific Partnership (TPP) agreement (see below). If TPP enters into force, most ICT tariff lines would fall to zero for TPP partner countries. This would include the aforementioned radio equipment tariffs imposed on U.S. exporters by Vietnam, which would fall to zero by Year 4 of TPP's implementation.[49]

Nontariff Barriers

Nontariff barriers (NTBs) are not as easily quantifiable as tariffs, but can also create significant hurdles to companies seeking to do business abroad. NTBs often come in the form of laws or regulations that intentionally or unintentionally discriminate and/or hamper the free flow of digital trade.

Nondiscrimination between local and foreign suppliers is a core principle encompassed in global trading rules and U.S. free trade agreements. While WTO agreements cover physical goods, services, and intellectual property, there is no explicit provision for nondiscrimination for digital goods. As such, NTBs that do not treat digital goods the same as physical ones could limit a provider's ability to enter a market.

Broader governance issues, including rule of law, transparency, and investor protections, can pose barriers and limit the ability for firms and individuals to successfully engage in digital trade.

Potential Barriers to Digital Trade

- High tariffs
- Localization requirements
- Cross border data flow limitations
- IPR infringement
- Discriminatory, unique standards or burdensome testing
- Filtering or blocking
- Cybertheft of U.S. trade secrets

Localization Requirements

Localization measures are defined as measures that compel companies to conduct certain digitaltrade-related activities within a country's borders.[50] Governments often use privacy or national security arguments as justifications for these measures. Though localization policies can be used to achieve legitimate public policy objectives, some are designed to protect, favor, or stimulate domestic industries, service providers, or intellectual property at the expense of foreign counterparts and, in doing so, function as nontariff barriers to market access. Free trade agreements, such as the TPP, aim to ensure an open Internet and eliminate trade barriers while preserving flexibility for governments to pursue legitimate policy objectives (see below).

Cross-Border Data Flow Restrictions

Regulations limiting cross-border data flows are a type of localization requirement that prohibit companies from exporting data outside a country. Such restrictions can pose barriers to companies whose transactions rely on the Internet to serve customers abroad and operate more efficiently. For example, data localization requirements can limit e-commerce transactions that depend on foreign financial service providers or multinational firms' full analysis of big data from across an entire company or global value chain. Regulations limiting cross-border data flows may force companies to build local server infrastructure within a country, not only increasing costs and decreasing scale, but also creating data silos that may be more vulnerable to cybersecurity risks.

Data localization requirements pose barriers to companies' efforts to operate more efficiently by migrating to the cloud. In 2014, 22% of businesses in OECD member countries used cloud computing services, with higher use among large enterprises, and the number is accelerating.[51] For example, AT&T has said that it plans to move 80% of its applications into a private cloud by the end of 2016.[52] To better serve consumers of Google's many cloud services (e.g., Gmail, search, maps) globally, the company is opening more data centers in the United States and internationally.[53] For companies more hesitant to embrace the cloud due to the security and regulatory concerns, Oracle Corp. has launched a hybrid cloud service offering.[54]

The Internet, and cloud services specifically, has been called the great equalizer as it allows small companies access to the same information and the same computing power as large firms using a flexible, scalable, and on-demand model. For example, Thomas Publishing Co., a U.S. mid-sized, private, family-owned and operated business, is transporting data from its own computer servers to data centers run by Amazon.com Inc.[55] A similar argument has been made for firms and governments in low and middle income countries who can take advantage of the power of the Internet to foster economic development.

> According to a USITC April 2015 report, the United States has the largest cloud computing industry globally (based on revenues) and 9 of the 10 largest cloud computing service providers (based on estimated number of servers).

Nevertheless, regulations or policies that limit data flows create barriers to firms and countries seeking to consume cloud services. As part of its submission to the U.S. Trade Representative (USTR) for the *2016 National Trade Estimate Report on Foreign Trade Barriers (NTE)*, for example, the Information Technology Industry Council (ITI) noted an increase in the use of forced localization measures, citing examples in China, Indonesia, Nigeria, Russia, Turkey and Vietnam.[56] The Business Software Alliance's 2016 Global Cloud Computing Scorecard highlighted countries with improved policy environments but also those with localization requirements, particularly Russia's data protection framework (which contains prescriptive data localization requirements).[57]

Other Localization Requirements

In addition to cross-border data flow restrictions, localization policies include requirements to use local content, whether hardware or software, as a condition for manufacturing or access to government procurement contracts; use local infrastructure or computing facilities; or partner with a local company and transfer technology or intellectual property to that partner. Localization requirements can also pose a threat to intellectual property (discussed below).

> **Examples of Localization Barriers**
>
> Examples cited in 2016 National Trade Estimate Report on Foreign Trade Barriers (NTE):[58]
>
> - In Turkey, a draft Personal Data Protection law would bar e-payment companies from the Turkish market if they do not have personal data banks located in Turkey.
> - In Nigeria, the government issued guidelines for ICT products requiring multinational companies in Nigeria to source all hardware locally; use only locally-manufactured SIM cards for telephone services and data; and use indigenous companies to build cell towers and base stations. The guidelines also require all government agencies to source and procure all computer hardware from government-approved original equipment manufacturers.
> - In India, the 2015 National Telecom M2M ("machine to machine") roadmap recommends preferences for locally-manufactured SIM cards and domestically-sourced goods, and requirements that application servers and gateways that serve customers in India be located domestically.

Intellectual Property Rights (IPR) Infringement

Intellectual property rights (IPR)[59] are legal, private, enforceable rights that governments grant to inventors and artists; they generally provide right holders with time-limited monopolies over the use of their creations, enabling them to exclude others from using their creations without their permission. IPR come in a variety of forms, such as patents, copyrights, trademarks, and trade secrets. While they are intended to encourage innovation and creative output by allowing inventors and artists to reap the benefits of the time and money they direct to developing IP, the rights are time-limited so that other inventors and artists can build on them and society can benefit more broadly through wider availability of works.

A wide range of U.S. industries rely on IPR protection. According to a 2012 report by the Department of Commerce, in 2010, IP-intensive industries accounted for about $5.06 trillion in value added, or 34.8% of U.S. gross domestic product (GDP).[60] These industries also were estimated to account for $775 billion (or 60.7% of) U.S. merchandise exports in 2010; and $90 billion (or nearly one-fifth) U.S. private services exports in 2007.[61] In 2014, U.S. charges for the use of IP (a U.S. services export) totaled $130.4 billion, while U.S. payments for the use of IP (a U.S. services import) totaled $42.1 billion, yielding a U.S. IP trade surplus of $88.3 billion.[62]

How Much IPR Infringement?

By its nature, IPR infringement is difficult to quantify, and estimates of its level and cost are sensitive to the assumptions made. Quantifying IPR infringement in the digital environment is all the more challenging given, for example, that "infringing files are traded online and websites offering counterfeits are launched and accessed, countless times each day."[63] By one estimate, nearly a quarter of global Internet traffic infringes on copyrights.[64] Another study pegs the value of digitally pirated music, movies, and software (not actual losses) as growing anywhere in the range of $80-$240 billion by 2015, up from $30-$75 billion in 2010.[65] Online sales of pirated and counterfeit goods reportedly could exceed the volume of sales "through traditional channels such as street vendors and other physical markets."[66] As a reference point, trade in physically traded counterfeit and pirated goods was estimated to be up to 2.5% ($461 billion) of world trade in 2013, based on custom seizures data (i.e., for goods seized at the border by countries' customs administrations).[67]

While the Internet and digital technologies have opened up and enhanced markets for international trade, they also have been a major driver of IPR infringement (e.g., theft of IP, such as through copyright piracy or counterfeiting of trademarks). Innovations in digital technologies fuel IPR infringement by enabling the rapid duplication and distribution of content that is low-cost and high-quality. For instance, IPR infringement makes it easy to pirate music, movies, software, and other copyrighted works and to share them globally. Newer innovations, like 3-D printing, could further facilitate IPR infringement by eliminating the need for a traditional factory to produce infringing goods because "all that will be needed is an Internet connection, a computer, and a 3-D printer."[68] The Internet provides "ease of conducting commerce through unverified vendors, inability for consumers to inspect goods prior to purchase, and deceptive marketing."[69]

Efforts to address IPR infringement raise issues of balance about, on one hand, protecting and enforcing IPR to incentivize innovation and, on the other hand, setting appropriate limitations and exceptions to ensure other economically and socially valuable uses. U.S. stakeholders differ on how to address such issues. Representatives of "content" industries have singled out Internet-enabled piracy as the most important barrier to digital trade for their industries (see *text box*). Barriers include foreign websites that facilitate IPR infringement, such as through hosting pirated content or connecting users to such content. Cyber theft of trade secrets presents additional, increasingly prominent, barriers to digital trade.[70] Content industries say that IP theft costs them sales, takes away from legitimate services, harms investors in these businesses, damages their brand or reputation, and hurts "law-abiding" consumers.[71]

**Examples of IPR Infringement
in Digital Trade**

- *Foreign websites that facilitate IPR infringement.* Some foreign websites offer large platforms to distribute globally infringing content (e.g., unauthorized copies of music, movies, software, video games) and illicit physical goods (e.g., counterfeit drugs). These websites take a variety of forms, including auction, business-tobusiness, consumer-to-consumer, and business-to-consumer sites. Some operate as "hubs" that allow users to upload content to file-sharing websites ("cyberlockers"), search applications that connect to websites to access content illegally (such as "e-libraries"), streaming sites that provide unauthorized access to copyrighted materials (such as "camcorded" copies of movies, and retransmission of live sports programs), and "pirate servers" that allow users to run unauthorized versions of cloud-based software. Countries in which parties host or operate online markets believed to be involved in commercial-scale IPR infringement include Brazil, Canada, China, Russia, Switzerland, and Ukraine.
- *Software piracy.* Issues include "end-user" piracy of software (e.g., installing software on multiple computers beyond license terms) and unauthorized installation of software, movies, music, and other creative programming. The use of illegal software by foreign governments is a particular concern.

- *Circumvention of technological protection measures (TPMs).* Measures such as encryption intended to limit the unauthorized reproduction, transmission, and use of products. Development and online distribution of devices that allow for TPM circumvention (e.g., modchips that allow users to play pirated games on physical consoles) raise IPR concerns.
- *Cybertheft of trade secrets.* Theft of trade secrets, including through cybertheft (e.g., cyber intrusions and hacking), appears to be escalating. Trade secrets are essential to many businesses' operations and important assets, including those in ICT, services, biopharmaceuticals, manufacturing, and environmental technologies. China is a top concern in terms of cybertheft of trade secrets, but other countries, such as India, also present challenges. Key issues include gaps in these countries' trade secret laws and enforcement, including criminal penalties that are not sufficient to act as deterrents.
- *Trademark infringement related to domain names.* Lack of protection of trademarks against unauthorized uses under country code top level domain names (ccTLDs) and "cybersquatting" is a concern for IPR-based businesses, and is related to the loss of Internet traffic.

Sources: USTR, 2015 Special 301 Report, April 2015 (designates countries that do not offer "adequate and effective" IPR protection and enforcement on various "watch lists"); USTR, 2015 Notorious Markets List, December 2015. (identifies foreign websites operating as online markets reportedly involved in commercial-scale IPR infringement); and ITC, Digital Trade in the U.S. and Global Economies, Part 1, USITC Publication 4415, July 2013.

Identifying those responsible for IPR infringement is challenging. Companies in technology products and services sectors express concerns over unpredictable legal frameworks in foreign countries for online intermediary liability regarding infringing or illegal content transmitted over their systems. For example, they contend that foreign courts use outmoded Internet service provider (ISP) liability laws that impose substantial penalties on ISPs, which deter investment and market entry and, in turn, impede legitimate online services.[72] Countries identified by the USTR as having imposed liability in ways that are contrary to U.S. intermediary liability policy include France, Germany, Italy, India, and Vietnam.[73]

Some technology product and service companies, as well as some civil society groups, also assert that overly stringent IPR policies may stifle information flows and legitimate digital trade. Thus, they highlight and promote exceptions and limitations to IPR, such as for "fair use"—a doctrine recognized in U.S. law that permits limited use of copyrighted works without requiring permission from the right holder in certain cases, such as criticism, comment, news reporting, research, scholarship, and teaching.[74] U.S. technology businesses voiced concern over proposals being considered or implemented in the European Union (EU) to charge U.S. and other providers of online search, news, and social media platforms a "quotation or snippet tax" for the "privilege" of quoting news publications; U.S. technology groups argue that this is contrary to international obligations.[75]

The USTR's *National Trade Estimate Report* similarly cites concerns regarding proposals for mandatory fees in the EU for linking to content published online, efforts that the USTR says appear to be targeting particular news aggregators that "index and allow users to more conveniently find and access such content by the inclusion in search results of headlines or other extracts of the stories that the underlying publisher typically offers, without charge (e.g., supported by advertising) on its own website."[76]

Other IPR-related barriers to digital trade include government measures, policies, and practices that are intended to promote domestic "indigenous innovation" (i.e., develop, commercialize, and purchase domestic products and technologies) but that can also disadvantage foreign companies. These measures can be linked to "forced" localization barriers to trade. China, for instance, conditions market access, government procurement, and the receipt of certain preferences or benefits on a firm's ability to show that certain IPR is developed in China or is owned by or licensed to a Chinese party. Another example is India's data and server localization requirements, which U.S. ITC firms assert hurts market access and innovation in their sector. (See above.)

National Standards and Burdensome Conformity Assessment

Local or national standards that deviate significantly from recognized international standards may make it difficult for firms to enter a particular market. An ICT product that conforms to international standards, for example, may not be able to connect to a local network or device based on a local or proprietary standard. Also, proprietary standards can limit a firm's ability to serve a market if their company practices or assets do not conform with (nor do their personnel have training in) those standards. As a result, customers in those markets have trouble accessing international providers.

Similarly, redundant or burdensome conformity assessment or local registration and testing requirements often add time and expense for a company trying to enter a new market, and serve as a deterrent to foreign companies. If a company is required to provide the source code for ICT products to gain market access, it may fear theft of their IP and not enter that market (see above).

Filtering, Blocking, and Net Neutrality

Governments that filter or block websites, or otherwise impede access, form another type of non-tariff barrier. For example, China has asserted a desire for "digital sovereignty" and has erected what is termed by some as the "great firewall." A recent change to China's Internet filters also blocks virtual private network (or VPN) access to sites beyond the great firewall. Virtual private networks have been used by Chinese citizens to use websites like Facebook.[77] A rule issued by China's State Administration of Press, Publication, Radio, Film and Television and the Ministry of Industry and Information Technology bans all foreign media from publishing online. According to press reports, apart from select individual collaboration projects, only companies that are 100% Chinese-owned will be able to produce online content, and only after approval from Chinese authorities and the acquisition of an online publishing license; foreign-owned or joint venture companies will be blocked from participating.[78]

According to recent press reports, Russia is now looking to emulate many of China's restrictive Internet practices.[79]

Due to the global nature of the Internet, one nation's preferences or regulations can have spillover effects on the rest of the world. French privacy authorities, for example, fined Google $112,000 for not applying a ruling on the "right to be forgotten" across the company's domains worldwide.[80] While Google had adopted the ruling by the Court of Justice of the European Union (CJEU) across all of its European operations, it had not done so globally, given that there is no one international standard or policy it is required to comply with. In one critic's view, "France is trying to force its domestic policies on the rest of the world by coercing a global company that resides in its borders to implement those policies on all its users."[81] The conflict between Google and the EU authorities illustrate the complexity of the Internet and evolving technologies, and the lack of global standards that prevails in other areas of international trade.

National-level neutrality policies also differ *widely*. *Net* neutrality rules govern the management of Internet traffic as it passes over broadband Internet access services (BIAS), whether those services are fixed or wireless. In contrast to China, in the United States, the Federal Communications Commission (FCC) rules ban the blocking of legal content, forbid paid prioritization of content for consideration or to benefit an affiliate, and prohibit the throttling of legal content by BIAS providers.[82] In the EU, however, the Telecoms Single Market legislation allows providers to offer a zero rating and have discretion on managing traffic during times of network congestion, subject to regulator's approval.[83] As a result, each end user's access may be subject to the preferences and decisions of a telecom supplier.

Cybersecurity Risks

The growth in digital trade has raised issues related to cybersecurity, the act of protecting ICT systems and their contents from cyberattacks. Cyberattacks in general are deliberate attempts by unauthorized persons to access ICT systems, usually with the goal of theft, disruption, damage, or other unlawful actions. Cybersecurity can also be an important tool in protecting privacy and preventing unauthorized surveillance or intelligence gathering. [84]

Cyberattacks can pose broad risks to financial and communication systems, national security, privacy, and digital trade and commerce. Examples in the commercial sector include hacks into JPMorgan's systems in which customers' personal information was accessed, later blamed on Iran, and breaches of Sony Pictures Entertainment in which proprietary information and internal communications were stolen and exposed, later blamed on North Korea.[85] Another issue is that companies that rely on cloud services to store or transmit data may choose to use enhanced encryption to protect the communication and privacy, both internally and of their end customers. This, in turn, may impede law enforcement investigations if they are unable to access the encrypted data.[86]

U.S. DIGITAL TRADE WITH THE EU AND CHINA

The European Union (EU) and China are large U.S. digital trade partners and each has presented various challenges for U.S. companies, consumers, and policymakers.

European Union

Differences in United States and EU policies have had ramifications on digital flows and international trade. The two partners' varying approaches to digital trade, privacy, and national security, have, at times, threatened to disrupt U.S.-EU data flows.

The transatlantic economy is the largest in the world, encompassing 46% of global GDP and 11% of the world's population.[87] Similarly, cross-border data flows between the United States and EU are the highest in the world. One estimate indicates that the United States exported $140.6 billion of digitally-delivered services to the EU in 2012, which was 72% of total U.S. exports to the EU.[88] Many of these services are used to create further exports as part of GVCs. Of the digitally-delivered services exported to the EU, 53% were incorporated into EU exports. In the opposite direction, the United States imported $86.3 billion of the same from the EU, 62% of which were incorporated into U.S. exports.[89] Furthermore, almost 40% of the data flows between the United States and EU are through business and research networks.[90]

Despite close economic ties, differences between the United States and EU in their approaches to data protection have caused friction in U.S.-EU economic and security relations. As of 2000, U.S. companies could use the Safe Harbor Agreement, negotiated between the United States and the EU, to transfer data across the Atlantic and comply with the EU-wide privacy framework established by the EU's 1995 Data Protection Directive (DPD). Nevertheless, in October 2015, the Court of Justice of the European Union issued a decision that invalidated Safe Harbor, finding that Safe Harbor did not provide "adequate" protection for personal data as required by EU law, in large part because of the U.S. surveillance programs disclosed in mid-2013.

In early 2016, U.S. and EU officials announced an agreement on a replacement to Safe Harbor— the EU-U.S. Privacy Shield which was approved by the European Commission (the EU's executive) and entered into force July 12, 2016.[91] While companies will be able rely on it to ensure their digital data flows are allowed, many experts expect that privacy advocates will challenge the Privacy Shield in court as well.[92]

The president of the French data protection authority (CNIL), Isabelle Falque-Pierrotin, who chairs the EU working party evaluating the agreement has stated that Privacy Shield is an opportunity to "build a common standard" between the EU and United States in cross-border data protection.[93] If the United States and EU are able to build a common standard, other parties may decide to adopt it, establishing a de facto global standard.

EU Digital Regulations

As illustrated with the DPD (discussed above), EU policymakers are attempting to bring more harmonization across the region. Another initiative is the Digital Single Market (DSM). The DSM is an ongoing effort to unify the EU market, facilitate trade, and drive economic growth. The DSM has three pillars: (1) better online access to digital goods and services through cross-border online activity; (2) high-speed, secure, trustworthy infrastructure that is supported by a regulatory environment supporting investment and fair competition; and (3) ensuring the digital economy as a driver for growth through investment in infrastructure, research and innovation, and an inclusive society and skilled citizen. The European Commission's strategy for a digital single market encompasses issues such as the portability of legally acquired content, cross-border data flows, copyright protection exceptions and limitations, intermediary liability, and enforcement.

Some voice concern about the extent to which the finalized DSM regulations will be consistent with U.S. companies' interests. For example, the United States has identified as a concern the Commission's consideration of a "duty of care" proposal as part of the DSM, which would "require some platforms to more proactively monitor and filter illegal content... despite logistical difficulties and implications for free expression."[94] Concerns also arise from the Commission's May 2016 package of e-commerce proposals that contain an update of the Audiovisual Media Services Directive (AVMSD) that includes rules on platform liability and local content requirements.[95]

While the DPD set out common rules on how information about European citizens may be collected and used across all industries, each EU member state is responsible for implementing the Directive through its own national laws. To modernize the DPD and facilitate the creation of the DSM, EU member states (acting in the Council of the European Union) and the European Parliament reached political agreement in late 2015 on a new

> General Data Protection Regulation (GDPR).[96] In contrast to the DPD, the GDPR will be directly applicable in all EU member states, thus establishing a single set of rules (rather than harmonized ones) for data protection throughout the EU. However, one observer contends that there are still approximately 40 provisions that allow individual member states to set their own standards.[97]
>
> The EU published the final GDPR on May 4, 2016; member states will have until May 25, 2018 to fully implement its provisions.[98] A dearth of guidance documents has caused U.S. industry to voice concern about the lack of clarity regarding some of the GDPR requirements and also about the potentially high penalties that may be imposed for violations (up to 2% of their annual worldwide revenues). Despite the lack of precise guidance, many companies have begun to analyze the regulation and plan for implementation. The potential impact of the GDPR on the EU-U.S. Privacy Shield is unclear, while the impact of the UK leaving the EU on either EU initiative is uncertain, although the UK's Information Commissioner supports amending UK data protection laws to meet the GDPR standards.[99]

These issues are likely to come up during the Transatlantic Trade and Investment Partnership (T-TIP) negotiations between the United States and EU. (See below.)

China

China presents a number of significant opportunities and challenges for the United States vis-a-vis digital trade. According to the Chinese government, at the end of December 2015, there were 688 million Internet users in China, including 620 million mobile Internet users. E-Marketer, a research firm that tracks digital issues, estimated China's e-commerce sales in 2014 totaled $672 billion (nearly double the U.S. level) and projected this would surge to $1.6 trillion by 2018.[100] Although many U.S. firms may benefit from expanding digital trade in China, they may face numerous challenges as well.

Internet Governance

In December 2015, Chinese President Xi Jinping in a speech declared that the international community should respect the Internet sovereignty of individual countries in "choosing their own Internet development path, Internet governance, and Internet policies." To many observers, this represents a growing effort by the Chinese government to expand its control over the Internet in China in a way that could have negative consequences for U.S. firms attempting to do business in China, as well as for Chinese entrepreneurs.

The USTR's 2016 National Trade Estimates of Foreign Trade Barriers stated: "Over the past decade, China's filtering of cross-border Internet traffic has posed a significant burden to foreign suppliers, hurting both Internet sites themselves, and users who often depend on them for their businesses."

Outright blocking of websites appears to have worsened over the past year, with 8 of the top 25 most trafficked global sites now blocked in China. Examples of blocked sites include Google services (e.g., Gmail), Twitter, Facebook, YouTube, and the New York Times. An example of the unpredictability of China's Internet market occurred in April 2016, when Chinese regulators, for unexplained reasons, suspended Apple iTunes Movies and iBooks Store, and DisneyLife services that had been operating in China for months.

IP Theft

China is considered by most analysts to be the largest source of global theft of IP. A May 2013 report by the Commission on the Theft of American Intellectual Property estimated that IP theft by Chinese entities annually cost the U.S. economy up to $240 billion. China is also considered to a major source of cyber-theft of U.S. trade secrets, including by government entities. In May 2014, the United States Department of Justice indicted five members of the Chinese People's Liberation Army (PLA) for government-sponsored cyber espionage against U.S. companies and theft of proprietary information to aid state-owned enterprises (SOEs). In April 2015, President Obama issued an executive order authorizing certain sanctions against "persons engaging in significant malicious cyber-enabled activates."

Shortly before the arrival of Chinese President Xi's state visit to the United States, in September 2015, the Obama Administration indicated that it was considering imposing sanctions against Chinese entities over cyber-theft, a move that likely could have led to a cancellation of Xi's visit. China sent a large delegation to the United States to discuss the issue, and during Xi's visit, the two sides reached an agreement whereby the two sides stated that "neither country's government will conduct or knowingly support cyber-enabled theft of intellectual property, including trade secrets or other confidential business information, with the intent of providing competitive advantages to companies or commercial sectors." The two sides also agreed to regularly hold high-level consultations on cyber issues.

ICT Policies

Many foreign companies have expressed concerns over the past few years over announced plans by the Chinese government to enact new national security, banking, and counterterrorism laws and regulations, including provisions that indicated the goal of having "secure and controllable" ICT. Many foreign ICT firms contend that the proposed rules were discriminatory and could be used to block them from the Chinese ICT market, or could require them to turn over sensitive technologies and intellectual property to the Chinese government.

During the 2015 U.S-China Strategic and Economic Dialogue (S&ED), China pledged to ensure that new ICT regulations would be nondiscriminatory and not impose nationality-based conditions or restrictions on foreign firms, re-affirming these commitments in the 2016 session. Both sides stated that they were committed to ensuring that ICT cybersecurity regulations "should be consistent with WTO agreements, be narrowly tailored, take into account international norms, be nondiscriminatory, and not impose nationality-based conditions or restrictions on the purchase, sale or use of ICT products by commercial enterprises unnecessarily," and that ICT cybersecurity measures generally applicable to the commercial sector would not "unnecessarily limit or prevent commercial sales opportunities for foreign suppliers of ICT products or services."[101]

In addition, over the past few years the Chinese government has increasingly emphasized the goal of boosting the level of innovation and technology development in China. This has led to discriminatory policies against foreign firms, such as "indigenous innovation" regulations that give preferences (such as for government procurement) to firms that use intellectual property developed in China. Other examples of discriminatory policies against foreign firms include restrictions on investment in telecommunications services, electronic payment services, and cloud computing. Additionally, U.S. ICT firms face a regulatory regime that is often non-transparent and unpredictable. In many instances, U.S. firms can only gain market access through joint ventures with a Chinese partner.

U.S.-China BIT Negotiations

In 2008, the United States and China launched negotiations for a bilateral investment treaty (BIT), an agreement that typically contains provisions to encourage and provide reciprocal investment protections in order to enhance bilateral commercial ties. In 2013, China agreed to negotiate a "high standard" BIT with the United States, which would include opening new sectors to FDI and generally treating U.S.-invested firms in China the same as Chinese firms. China agreed to negotiate investment liberalization on a negative list basis, meaning only those industries listed in the agreement would be closed off to foreign investment—all other sectors would be open. Many analysts contend that a BIT could significantly boost bilateral FDI and trade flows. Such an agreement, if concluded, might provide significant new opportunities for U.S. firms that are engaged in digital trade.[102]

U.S.-China Cybersecurity Working Group

As a result of the 2015 S&ED meeting and cybersecurity agreement, the United States and China established U.S.- China High-Level Joint Dialogue on Cybercrime and Related Issues.[103] According to the White House, the group "will be used to review the timeliness and quality of responses to requests for information and assistance with respect to malicious cyber activity of concern identified by either side." The group first met in December 2015 and agreed on guidelines, conducting a tabletop exercise, a hotline mechanism, and enhanced cooperation on cyber-enabled crime.[104]

At the second meeting in June 2016, the parties agreed to a second tabletop exercise, implementation of the hotline, cooperation in network protection, information sharing, and the first U.S.-China Senior Experts Group on International Norms in Cyberspace and Related Issues. They also agreed to cooperate on investigations, combatting IP theft, and law enforcement operations in specific areas.[105]

As negotiations with each the EU and China demonstrate, there no single international standard governs digital data flows, and the topic is treated inconsistently, if at all, in trade agreements.

A United National Conference on Trade and Development (UNCTAD) report exploring data protection pointed out that differences in social and cultural norms affect if, and how, countries regulate privacy, which in turn can have trade implications.[106] In reviewing privacy and data flow regimes at national and regional levels globally, UNCTAD identified common core principles: openness, collection limitation, purpose specification, use limitation, security, data quality, access and correction, accountability.[107] The report urges global work toward an agreement or mechanism to promote international harmonization or compatibility between the different regimes. After all, "(c)reating trust online is a fundamental challenge to ensuring that the opportunities emerging in the information economy can be fully leveraged."[108]

Despite common core principles, governments face multiple challenges in designing policies. The OECD points out three potentially conflicting policy goals in the Internet economy: (1) enabling the Internet; (2) boosting or preserving competition within and outside the Internet; and (3) protecting privacy and consumers more generally.[109]

DIGITAL TRADE PROVISIONS IN TRADE AGREEMENTS

As digital trade has emerged as an important component of trade flows, it has risen in significance on the trade policy agenda of many countries, including the United States. Given the current stalemate in the WTO Doha Round negotiations, multilateral trade agreements have not kept pace with the complexities of the digital economy and digital trade is treated unevenly, if at all, in existing WTO agreements. More recent bilateral and plurilateral deals have started to address digital trade more comprehensively. The use of digital trade provisions in bilateral and plurilateral trade negotiations may help spur interest in the creation of future WTO frameworks that focus on digital trade.

WTO Provisions

While no comprehensive agreement on digital trade exists in the WTO, other WTO agreements cover some aspects of digital trade.

General Agreement on Trade in Services (GATS)

The WTO General Agreement on Trade in Services (GATS) entered into force in January 1995, predating the current reach of the Internet and the explosive growth of global data flows. GATS includes obligations on nondiscrimination and transparency that cover all service sectors. The market access obligations under GATS, however, are on a "positive list" basis in which each party must specifically opt in for a given service sector to be covered.[110]

As GATS does not distinguish between means of delivery, trade in services via electronic means is covered under GATS. While GATS contains explicit commitments for telecommunications and financial services that underlie e-commerce, digital trade and information flows and other trade barriers are not specifically included. Given the positive list approach of GATS, coverage across members varies and many newer digital products and services did not exist when the agreements were negotiated.

Addressing new topics like e-commerce and data flows has been raised but not yet formalized in the WTO. The 10th Ministerial Conference of the WTO, in December 2015, concluded with no clear path forward for the Doha Development Agenda (DDA), reflecting an ongoing wide division among members. Most developing countries want to continue the DDA round that links the broad spectrum of agricultural and nonagricultural issues, maintaining that unless all issues are addressed in a single package, issues important to developing countries will be ignored.

Conversely, advanced economies, including the United States and EU, have pushed for an end to the long-stalled round, arguing that the Doha agenda has proven untenable and that a different approach is needed in order to address new issues including e-commerce and data flows. While members claim to remain committed to addressing the outstanding issues of the round, both agricultural and nonagricultural, the Nairobi Ministerial Declaration acknowledged the division over the future of the Doha Round, and failed to reaffirm its continuation, leaving its future uncertain.[111]

Information Technology Agreement (WTO ITA)

The World Trade Organization (WTO) Information Technology Agreement (ITA) aims to eliminate tariffs on the goods that power and utilize the Internet. Originally concluded in 1996, the ITA was expanded during the WTO's Tenth Ministerial Conference in December 2015, entering into force in July 2016. The expanded ITA is a plurilateral agreement among 54 developed and developing WTO members who account for over 90% of global trade in these goods. Some WTO members, such as Vietnam and India, are party to the original ITA, but did not join the expanded agreement. Like the original ITA, the benefits of the expanded agreement will be extended on a most-favored nation (MFN) basis to all WTO members.

The expanded ITA will eliminate tariffs on 201 additional IT products valued at over $1.3 trillion per year.[112] The increased coverage includes, for example, many consumer electronics, new generation semi-conductors (multi-component semiconductors, or MCOs), and medical instruments like magnetic resonance imaging (MRI). According to the U.S. Trade Representative (USTR), the agreement will provide duty-free access to $180 billion in annual U.S. exports.[113] The parties also agreed to review the agreement's scope no later than 2018 to determine if additional product coverage is warranted as technology evolves.

While the WTO ITA is expected to expand trade in the technology products that underlie digital trade, it does not tackle the nontariff barriers that can pose significant limitations.

Declaration on Global Electronic Commerce

In May 1998, WTO members established the "comprehensive" Work Programme on Electronic Commerce "to examine all trade-related issues relating to global electronic commerce, taking into account the economic, financial, and development needs of developing countries."[114] The 1998 declaration establishing the program also included a statement that "members will continue their current practice of not imposing customs duties on electronic transmission."[115]

Reflecting the lack of agreement in the final WTO Ministerial Declaration, the latest report for the work program stated that there was not consensus on how to move forward beyond the information sharing stage to identify specific outcomes or recommendations.[116] In the draft decision in November 2015, members agreed to continue periodic reviews of the work program, the current moratorium on customs duties on electronic transmissions, and having the other WTO bodies explore the relationship between existing WTO agreements and e-commerce based on proposals submitted by members.[117]

Trade-Related Aspects of Intellectual Property Rights (TRIPS)

The TRIPS Agreement, signed on April 15, 1994, and in effect since January 1, 1995, provides minimum standards of IPR protection and enforcement. The TRIPS Agreement does not specifically cover IPR protection and enforcement in the digital environment, but arguably has application to the digital environment and sets a foundation for IPR provisions in subsequent U.S. trade negotiations and agreements, many of which are "TRIPS-plus."

The TRIPS Agreement covers copyrights and related rights (i.e., for performers, producers of sound recordings, and broadcasting organizations), trademarks, patents, trade secrets (as part of the category of "undisclosed information"), and other forms of IP. It builds on international IPR treaties, dating to the 1800s, administered by the World Intellectual Property Organization, or WIPO (see below). TRIPS incorporates the main substantive provisions of WIPO conventions by reference, making them obligations under TRIPS. WTO members were required to fully implement TRIPS by 1996, with exceptions for developing country members by 2000 and leastdeveloped-country (LDC) members until July 1, 2021, for full implementation.[118]

TRIPS aims to balance rights and obligations between protecting private right holders' interests and securing broader public benefits. It includes provisions on:

- WTO nondiscrimination principles (national treatment and most-favored-nation);
- Minimum standards of protection for IPR, such as copyright protection terms for the life of the author plus 50 years;
- Minimum standards of enforcement of IPR through civil actions for infringement, border enforcement, and criminal actions;
- Applying the WTO's binding Dispute Settlement Mechanism to IPR disputes; and,
- Requiring developed countries to provide incentives for technology transfers to LDCs "to enable them to create a sound and viable technological base."

Among other provisions, the TRIPS section on copyright and related rights includes specific provisions on computer programs and compilations of data. It requires protections for computer programs—whether in source or object code—as literary works under the WIPO Berne Convention for the Protection of Literary and Artistic Works (Berne Convention). TRIPS also clarifies that databases and other compilations of data or other material, whether in machine readable form or not, are eligible for copyright protection even when the databases include data not under copyright protection.[119]

Like the GATS, TRIPS predates the era of ubiquitous Internet access and commercially significant e-commerce. TRIPS includes a provision for WTO members to "undertake reviews in the light of any relevant new developments which might warrant modification or amendment" of the agreement. The TRIPS Council has engaged in discussions on the agreement's relationship to electronic commerce as part of the WTO Work Programme on Electronic Commerce, focusing on protection and enforcement of copyright and related rights, trademarks, and new technologies and access to these technologies.[120]

World Intellectual Property Organization (WIPO) Internet Treaties

The World Intellectual Property Organization (WIPO) has been a primary forum to address IP issues brought on by the digital environment since the TRIPS Agreement. The WIPO Copyright Treaty and WIPO Performances and

Phonograms Treaty—often referred to jointly as the WIPO "Internet Treaties"— established international norms regarding IPR protection in the digital environment. These treaties were agreed to in 1996 and entered into force in 2002, but are not enforceable under WTO dispute settlement. Shaped by TRIPS, the WIPO Internet Treaties are intended to clarify that existing rights continue to apply in the digital environment, to create new online rights, and to maintain a fair balance between the owners of rights and the general public.[121]

Key features of the WIPO Internet Treaties include provisions for legal protection and remedies against circumventing TPMs, such as encryption, and against the removal or alteration of rights management information (RMI), which is data identifying works or their authors necessary for them to manage their rights (e.g., for licenses and royalties).

The liability of online service providers and other communication entities that provide access to the Internet was contested in the negotiations on the WIPO Internet Treaties. An "agreed statement" regarding Article 8 of the WIPO Copyright Treaty sought to clarify the issue by providing that "the mere provision of physical facilities for enabling or making a communication [e.g., wires, telephone lines, modems] does not in itself amount to communication within the meaning of this Treaty or the Berne Convention...." The WIPO Internet Treaties leave it to the discretion of national governments to develop the legal parameters for Internet Service Provider (ISP) liability.[122]

As of April 2016, the WIPO Internet Treaties had 94 parties. The United States implemented the WIPO Internet Treaties through the Digital Millennium Copyright Act of 1998 (DMCA), which set new standards for protecting copyrights in the digital environment, including prohibiting the circumvention of anti-piracy measures incorporated into copyrighted works and enforcing such violations through civil, administrative, and criminal remedies.[123] The DMCA also, among other things, limits remedies available against ISPs that unknowingly transmit copyright infringing information over their networks by creating certain "safe harbors."[124] The United States continues to calls on trading partners, such as Canada and Mexico, to fully implement the WIPO Internet Treaties.[125]

Future Sectoral Approaches

With the stalling of the Doha Round of negotiations, WTO members and experts have raised various options to address emerging issues such as digital trade. Ideas include:

- Updating the rules within the WTO framework to address digital trade.[126] Options could include expanding the multilateral GATS to cover cross-border data flows, technology transfer, or greater market access issues. Others support using the existing plurilateral WTO ITA, Telecommunications, or the Trade Facilitation Agreement to address digital trade and tackle barriers ranging from tariffs to express delivery and mobile services.
- Establishing a permanent WTO working group dedicated to exploring digital issues, possibly based on the current Work Programme, or to create new standalone trade agreement specific to data services or digital trade, possibly initially as an open plurilateral deal.
- Creating a separate digital trade-specific WTO agreement, a "e-WTO" as some have suggested. USTR Ambassador Froman noted that "[n]ew rules on critical 21st century issues, such as e-commerce and the digital economy, are emerging. ... a better path forward is a new form of pragmatic multilateralism. Moving beyond Doha doesn't mean leaving its unfinished business behind. Rather, it means bringing new approaches to the table."[127]

In July 2016, the United States put forward a submission under the WTO Work Programme on Electronic Commerce offering "trade-related policies that can contribute meaningfully to the flourishing of trade through electronic and digital means" but without specific negotiating proposals.[128] The 16 policies included in the U.S. submission align with the proposed Trans-Pacific Partnership (see below) such as, prohibiting digital customs duties and enabling cross-border data flows. The policies focus on eliminating or preventing trade barriers and establishing a transparent, adaptable framework for digital trade. The policies also recognize the need for balancing digital trade with other priorities such as protection of consumer data, security, and law enforcement.[129]

U.S. Bilateral and Plurilateral Agreements

As discussed above, the WTO agreements provide limited treatment of some aspects of digital trade. The stalled Doha Round and the desire by some parties to address new topics such as e-commerce are two of the drivers behind the growth of bilateral and plurilateral trade agreements outside of the WTO.

The United States has included, and continues to expand on, digital trade provisions in its bilateral and plurilateral trade negotiations.

Existing U.S. Free Trade Agreements (FTAs)

The United States has included an e-commerce chapter in its FTAs, since it signed an agreement with Singapore in 2003.[130] The e-commerce chapter of U.S. FTAs usually begins by recognizing e-commerce as an economic driver and the importance of removing trade barriers to ecommerce.[131] Most chapters contain provisions on nondiscrimination of digital products, prohibition of customs duties, transparency, and cooperation topics such as SMEs, cross-border information flows, and promoting dialogues to develop e-commerce. Some of the FTAs also include cooperation on consumer protection, as well as providing for electronic authentication and paperless trading. All FTAs allow certain exceptions to ensure that each party is able to achieve legitimate public policy objectives, protecting regulatory flexibility.

Electronic Commerce Chapter Article 1 in U.S. FTAs

"The Parties recognize the economic growth and opportunity that electronic commerce provides, the importance of avoiding barriers to its use and development, and the applicability of the WTO Agreement to measures affecting electronic commerce."

The U.S.-South Korea FTA (KORUS) contains the most robust digital trade provisions in a U.S. FTA currently in force.[132] In addition to the provisions in prior FTAs, KORUS includes provisions on access and use of the Internet to ensure consumer choice and market competition. Most significantly, KORUS was the first attempt in a U.S. FTA to explicitly address cross-border information flows. The e-commerce chapter contains an article that recognizes its importance and discourages the use of barriers to cross-border data but does not mention explicitly localization requirements. The financial services chapter of KORUS also contains a specific, enforceable commitment to allow cross-border data flows "for data processing where such processing is required in the institution's ordinary course of business."[133]

The Proposed Trans-Pacific Partnership (TPP) Agreement

The Trans-Pacific Partnership (TPP) is a proposed FTA among 12 Asia-Pacific countries, including both developed and developing countries. The agreement has economic and strategic significance for the United States and was officially signed on February 4, 2016. [134] Congress must pass implementing legislation before the TPP agreement can take effect in the United States. In considering TPP, Congress may weigh whether the agreement makes enough progress in achieving the TPA negotiating objectives on digital trade to merit passage of implementing legislation.

The proposed TPP goes beyond the digital trade provisions in KORUS and earlier U.S. FTAs. Overall, the agreement aims to promote digital trade, the free flow of information, and ensure an open Internet. Provisions related to digital trade are included in multiple chapters of the TPP (e.g., e-commerce, financial services, telecommunications, technical barriers to trade, intellectual property rights), showing the complexity of digital trade barriers and issues. The TPP encourages parties to become members of the tariff-eliminating WTO Information Technology Agreement. In reviewing the TPP, the Industry Trade Advisory Committee on Information and Communication Technologies Services and Electronic Commerce (ITAC 8) endorsed the agreement, finding that the TPP promotes the economic interests of the United States, and provides equity and reciprocity for the sectors represented by the ITAC.[135]

The proposed TPP has several digital trade-related innovations, including:

- Prohibits cross-border data flow restrictions and data localization requirements, except for financial services and government procurement.
- Prohibits requirements for source code disclosure or transfer as a condition for market access, with exceptions.
- Requires parties to have online consumer protection and anti-spam laws, and a legal framework on privacy.
- Prohibits requiring technology transfer or access to proprietary information for products using cryptography.
- Clarifies IPR enforcement rules to provide criminal penalties for trade secret cybertheft.
- Encourages cooperation between parties on e-commerce to assist SMEs, and on privacy and consumer protection.
- Promotes cooperation on cybersecurity.
- Safeguards cross-border electronic card payment services.
- Covers mobile service providers and promotes cooperation for international roaming charges.

In addition to excluding government procurement, TPP allows for exceptions to its digital trade commitments to achieve legitimate public policy goals such as protecting health, safety, and national security. Like other FTAs, the TPP also includes annexes of nonconforming measures in which each country negotiates to exclude specific regulations, laws, or sectors from its agreement obligations. For example, Japan includes national security screening requirements on "telecommunications and internet based services."[136] Unless a country takes an exception through a nonconforming measure, the "negative" list approach of TPP would ensure that new services or innovations would be covered under the agreement obligations.

For the first time, TPP would require parties to have a legal framework to protect personal information. TPP critics contend that the provisions are vague and do not contain an explicit minimum standard for privacy protection. Supporters note that TPP includes a reference to take into account "guidelines of relevant international bodies" that may include the Asia-Pacific Economic Cooperation (APEC) Privacy Framework.[137]

While most industry advocates support TPP, critics point out that financial services are not covered by the overall e-commerce chapter. The financial services chapter, instead, includes a separate provision covering cross-border data flows based on the language in KORUS, but it does not contain a prohibition on localization requirements similar to the e-commerce chapter.[138]

Trade in Services Agreement (TiSA) Negotiations

Negotiations on a proposed plurilateral Trade in Services Agreement (TiSA) were launched in April 2013, and are occurring outside of the WTO, with a goal of concluding the agreement in 2016.[139] The 23 TiSA participants account for about 70% of world trade in services and include the United States, EU, and Australia. Some key major emerging markets, including Brazil, China, and India, are not currently parties to the TiSA negotiations.

While nondiscrimination (MFN) applies to all services sectors, in TiSA, unlike TPP, market liberalization commitments are being negotiated under a hybrid approach. That is, specific market access obligations to liberalize service markets are being negotiated under a positive list in which parties "opt in" specific service sectors, while national treatment obligations are being negotiated under a negative list (in which parties may "opt out" certain sectors or sub-sectors). The positive list may be viewed as less ambitious because new inventions or sectoral innovations would not be covered under TiSA unless they are explicitly added in the future, a potential concern in the quickly evolving world of digital trade.

Though the final structure and sectors to be covered in TiSA remain under negotiation, setting common rules for digital trade is a key interest of the United States. The chapter or annex on digital trade or e-commerce would likely address trade barriers to cross-border data flows, consumer online protection, and interoperability, among other areas, similar to the provisions in the proposed TPP. [140]

The United States reportedly advocates applying TPP provisions on data localization and cross-border data flows to the entire TiSA agreement by placing these not solely in the e-commerce section, but rather in the core text as a horizontal obligation that would cover all sectors. If it is in the core text, no explicit commitment on data flows or localization in the e-commerce section may be needed, though parties could still choose to exclude a sector(s) through a nonconforming measure. Requiring regulatory cooperation and ongoing dialogue on digital trade issues between TiSA members could provide a path forward without changing existing laws in each TiSA country.

Negotiators could decide to include international regulatory cooperation on matters of cybersecurity or in support of small and mid-sized enterprises as in TPP. Negotiators may aim for language that is open enough to enable trade and address evolving technology, but concrete enough for regulators to protect privacy and safeguard cybersecurity.

Transatlantic Trade and Investment Partnership (T-TIP) Negotiations

T-TIP is a potential FTA that the United States and the EU have been negotiating since 2013 to reduce and eliminate tariff and nontariff barriers on goods, services, and agriculture, as well as to establish globally relevant trade rules and disciplines that expand on WTO commitments and address newer issues. Digital trade is a key area of interest because of its significance to transatlantic trade. Services that can be delivered over the Internet constitute the majority of U.S. and EU services exports to each other.[141]

T-TIP is expected to address digital trade issues, but may not include data privacy standards. The European Parliament's nonbinding T-TIP resolution calls for the European Commission to ensure that data privacy is not compromised in the liberalization of data flows while recognizing the importance of data to U.S.-EU trade and the digital economy.[142] Stakeholders differ on how T-TIP should address digital trade, particularly on how to balance promoting cross-border data flows against protecting data and ensuring data privacy. Many in a wide range of U.S. and EU industries support "horizontal and binding" commitments to promote the free flow of data across borders.[143] Others, including some consumer advocates, share the EU's concern about the potential implications for data protection and data privacy of including data flows in T-TIP.[144]

Potential T-TIP Provisions

The United States and EU aim to achieve a comprehensive and high-standard T-TIP, but continue to negotiate the final structure and scope of T-TIP. Potential T-TIP provisions include:

Market access. Provisions on e-commerce could provide enhanced market access for digital products. A U.S. negotiating objective is to develop "appropriate provisions to facilitate the use of electronic commerce to support goods and services trade, including through commitments not to impose customs duties on digital products or unjustifiably discriminate among products delivered electronically."[145] Some have called for a "negative list" approach to ensure that future innovations can be covered.[146]

Regulatory cooperation. The United States and the EU have "different legal traditions, regulatory paths, market outcomes," and policymaking approaches that constrain integration of a transatlantic digital economy. A focus for the United States is ensuring horizontal commitments, such as on notice and comment for stakeholder input, and transparency features of the regulatory processes. The United States and the EU also have identified ICT as one of the sectors for enhanced regulatory cooperation.

Rules on e-commerce. Rules to facilitate data flows across borders and address localization requirements (e.g., data storage or server location requirements) could promote e-commerce. Such rules could be horizontal (i.e., applying to all sectors) unless the United States or EU takes an exception to the obligation. Given that the proposed TPP e-commerce chapter excludes financial services, there may be particular interest in seeing how T-TIP approaches this issue.

IPR rules. Commitments to protect and enforce IPR, including copyrights, in the digital environment provide an opportunity to address digital trade issues. Such commitments could be balanced against limitations on ISP liability and "fair use" exceptions, could protect against forced transfers of source code, or establish criminal procedures for cyber theft.

Digital trade issues, among others, may be particularly contentious. Outcomes on this issue may be complicated or influenced by a number of factors, such as the EU's efforts to create a "Digital Single Market." Other developments that may shape T-TIP include new and revised EU policies on data protection, EU-U.S. Privacy Shield implementation, and TPP's outcome. (See "U.S. Digital Trade ".) The impact of the UK's decision to leave the EU is uncertain as the exact process, timing, and the UK's relationship with the EU post-membership remain unclear.

OTHER INTERNATIONAL FORUMS FOR DIGITAL TRADE

Given the cross-cutting nature of the digital world, digital trade issues touch on other policy objectives and priorities, such as privacy and national security. While U.S. and international trade agreements may be one way for the United States to instill firm obligations with trading partners, not every issue is necessarily suitable for an international trade agreement and not every international partner is ready, or willing, to take on such commitments. In other international forums outside of trade negotiations, other tools can be used to encourage high-level, non-binding best practices, principles, and align expectations.

G-20. The influential Group of 20 (G-20) is one venue for establishing common principles and digital issues have been on their agenda recently.[147] At their November 2015 meeting, the G-20 leaders issued a statement that included new provisions on the Internet economy, recognizing the opportunities and challenges presented to global economic growth and development, affirming not to conduct or support ICT-enabled IP theft for commercial competitive advantage, and acknowledging the need to respect and protect privacy.[148] China is the 2016 host for the G-20 and selected the theme "Towards an Innovative, Invigorated, Interconnected and Inclusive World Economy," opening the opportunity for further conversation on the digital economy and the chance to set global norms.

G-7. The G-7 ICT Ministers met in Japan in April 2016 and issued a Joint Declaration, stressing principles including the importance of investment in infrastructure, digital literacy, and accessibility; promoting cross-border data flows, privacy and data protection, and cybersecurity; and fostering innovation through open markets, interoperable standards, protecting IP, and facilitating research and development.[149] The United States could work with G-7 partners to incorporate these principles into the broader G-20.

OECD. The OECD offers yet another forum to discuss principles and norms to ensure a thriving digital economy. The June 2016 Ministerial Meeting in Mexico, titled "Digital Economy: Innovation, Growth and Social Prosperity," addressed an open Internet and data flows; infrastructure and connectivity; digital trust; and workforce skills.[150] The Ministerial Declaration included recognizing the growth and transforming impact of the digital economy as well as evolving challenges, and declared support of the free flow of information, innovation and emerging technologies, and the need to build trust, reduce impediments to e-commerce, and enable opportunities.[151] The declaration also acknowledged the need to balance public policy objectives and incorporate a whole-of-society perspective. The United States could work with OECD partners to reinforce these principles by defining specific action plans or commitments.

APEC. The Asian Pacific Economic Cooperation (APEC) forum presents another opportunity for sharing best practices and setting high level principles on issues that may be of greater concern to developing countries with less advanced digital economies and industry.[152] The APEC Electronic Commerce Steering Group (ECSG) coordinates e-commerce activities for APEC and promotes the development and use of e-commerce legal, regulatory and policy environments that are predictable, transparent and consistent. Within the ECSG, APEC is developing and implementing a Cross-Border Privacy Rules system to be consistent with the already established APEC Privacy Framework.[153] While APEC initiatives are regionally focused, because they reflect economies at different stages of development and include industry participation, they can provide a basis to scale up to larger global efforts. Due to its voluntary nature, APEC can serve as an incubator for potential plurilateral agreements. As such, by maintaining U.S. involvement in APEC, the United States can guide efforts to establish principles and norms in the region and subsequent roll-out worldwide.

Regulatory cooperation. Congress could consider having U.S. regulatory agencies that cover specific aspects of digital trade (e.g., U.S. Federal Trade Commission, Customs and Border Protection) work with overseas counterparts to better align regulatory requirements and reduce inconsistencies and redundancies that can hamper or discriminate against the free flow of data, goods, and services. Online privacy, consumer protection across borders, and rules for online contract formation and enforcement are potential areas for regulatory cooperation. The EU-U.S. Privacy Shield is one example of regulatory authorities working together to address such issues.

POLICY ISSUES FOR CONGRESS

Policy questions continue to evolve as the Internet-driven economy and innovations grow. Digital trade is intimately connected to and woven into all parts of the U.S. economy and overlaps with other sectors, requiring policymakers to balance many different objectives. For example, digital trade relies on cross-border data flows, but policymakers must balance open data flows with public policy goals such as protecting privacy, supporting law enforcement, and improving personal and national security and safety.

The complexity of the debate related to cross-border data flows involves complementary and competing interests and stakeholders. Companies and individuals who seek to do business abroad, and trade negotiators who seek to open markets, are concerned with maintaining open market access, which may include cross border data flows, while others may want to limit foreign competition. Privacy advocates focus on protecting personal information. Meanwhile, law enforcement and defense advisors may seek the ability to access or limit information flows based on national security interests.

Digital trade raises numerous complex issues of potential interest to Congress with potential legislative and oversight implications. Issues include:

- Understanding of the economic impact of digital trade on the U.S. economy and the effects of localization and other digital trade barriers on U.S. exports and competition.
- Examining how best to balance an individual's right to privacy for conduct online and the government's need for access to protect safety and national security.
- Considering how best to assure public confidence and trust in network reliability and security that underlie the global digital economy and allow it to effectively and efficiently function.
- Reviewing what government policies to pursue with the private sector to support innovation and economic growth in digital trade both domestically and internationally.
- Examining the evolving U.S. trade policy efforts including the EU-U.S. Privacy Shield, TPP, and WTO policy principles to determine if these mechanisms establish an appropriate balance amongst public policy objectives, and conducting oversight of implementation should they enter into force.
- Assessing if China is abiding by its commitments in the bilateral cyber agreement and on market access for U.S. ICT firms, as well as the effectiveness of the bilateral cyber dialogue.
- Reviewing federal-level efforts related to digital trade, such as the Department of Commerce's Digital Economy Agenda or

infrastructure programs, and determine if changes to current plans or funding levels are needed.
- Conducting oversight of federal agencies in terms of roles and competencies related to digital trade, such as those organizations charged with coordinating federal efforts on IPR or law enforcement; trade negotiations and enforcement; and cybersecurity.

End Notes

[1] James Manyika, et al, Digital globalization: The new era of global flows, February 2016, http://www.mckinsey.com/business-functions/mckinsey-digital/our-insights/digital-globalization-the-new-era-ofglobal-flows?cid=other-eml-alt-mgi-mck-oth-1602.

[2] Susan Ariel Aaronson, The Digital Trade Imbalance and Its Implications for Internet Governance , Centre for International Governance Innovation and Chatham House, 2016, p. 1, https://www.cigionline.org/sites/default/files/gcig_no25_web_0.pdf.

[3] U.S. International Trade Commission, Digital Trade in the U.S. and Global Economies, Part 2, Publication No: 4485, Investigation No: 332-540, p.29, August 2014, https://www.usitc.gov/publications/332/pub4485.pdf.

[4] Paul Zwillenberg, Dominic Field, and David Dean, Greasing the Wheels of the Internet Economy, Boston Consulting Group, February 2014, https://www.bcgperspectives.com/content/articles/digital_economy_telecommunications_greasing_wheels_internet_economy/.

[5] Business Software Alliance, Powering the Digital Economy: A Trade Agenda to Drive Growth, 2015, http://www.bsa.org/~/media/Files/Policy/Trade/DTA_study_en.pdf.

[6] The World Bank Group, World Development Report 2016: Digital Dividends, 2016, http://www.worldbank.org/en/publication/wdr2016.

[7] The United States was not included in the study. OECD. (2015), "Executive summary," OECD Digital Economy Outlook 2015, p. 2-3, OECD Publishing, Paris. DOI: http://dx.doi.org/10.1787/9789264232440-2-en.

[8] Internet Association, Measuring the U.S. Internet Sector, 2015, http://internetassociation.org/wpcontent/uploads/2015/12/Internet-Association-Measuring-the-US-Internet-Sector-12-10-15.pdf.

[9] Giulia McHenry, Evolving Technologies Change the Nature of Internet Use, National Telecommunications & Information Administration blog, April 19, 2016.

[10] Susan Lund and James Manyika, Strengthening the Global Trade and Investment System for Sustainable Development: How Digital Trade is Transforming Globalization., The E15 Initiative. McKinsey & Company., January 2016, http://e15initiative.org/publications/how-digital-trade-is-transforming-globalisation/.

[11] Erik van der Marel, Disentangling the Flows of Data: Inside or Outside the Multinational Company? European Center for International Political Economy, July 2015, http://ecipe.org/publications/flows-data-inside-outsidemultinational-company/?chapter=all.

[12] James Manyika, Jacques Bughin, and Susan Lund, et al. Global Flows in a digital age: How trade, finance, people, and data connect, McKinsey Global Institute, April 2014, http://www.mckinsey.com/business-functions/strategy-andcorporate-finance/our-insights/global-flows-in-a-digital-age.

[13] ICT is an umbrella term that includes any communication device or application, including: radio, television, cellular phones, computer and network hardware and software, satellite systems, and associated services and applications.

[14] Business Software Alliance, Powering the Digital Economy: A Trade Agenda to Drive Growth, January 2014, p.8-9, http://digitaltrade.bsa.org/pdfs/DTA_study_en.pdf.

[15] OECD. (2015), "Chapter 2: The foundations of the digital economy," OECD Digital Economy Outlook 2015, p. 92, OECD Publishing, Paris. DOI: http://dx.doi.org/10.1787/9789264232440-2-en.

[16] Susan Lund and James Manyika, Strengthening the Global Trade and Investment System for Sustainable Development: How Digital Trade is Transforming Globalization., The E15 Initiative. McKinsey & Company., January 2016, http://e15initiative.org/publications/how-digital-trade-is-transforming-globalisation/.

[17] The OECD defines the Internet of Things as "encompassing all devices and objects whose state can be read or altered via the internet, with or without the active involvement of individual... The internet of things consists of a series of components of equal importance – machine-to-machine communication, cloud computing, big data analysis, and sensors and actuators. Their combination, however, engenders machine learning, remote control, and eventually autonomous machines and systems, which will learn to adapt and optimise themselves." OECD (2015), OECD Digital Economy Outlook 2015, p. 61, OECD Publishing, Paris. DOI: http://dx.doi.org/10.1787/9789264232440-2-en. For more information on the Internet of Things, see CRS Report R44227, The Internet of Things: Frequently Asked Questions, by Eric A. Fischer.

[18] This is distinct from the physical infrastructure platforms for data and connectivity that these digital platforms rely on.

[19] According to the U.S. National Institute of Standards and Technology, cloud computing is a model for enabling ubiquitous, convenient, on-demand network access to a shared pool of configurable computing resources (e.g., networks, servers, storage, applications, and services) that can be rapidly provisioned and released with minimal management effort or service provider interaction. For more information, see CRS Report R42887, Overview and Issues for Implementation of the Federal Cloud Computing Initiative: Implications for Federal Information Technology Reform Management, by Patricia Moloney Figliola and Eric A. Fischer 20 The World Bank Group, World Development Report 2016: Digital Dividends, 2016, http://www.worldbank.org/en/publication/wdr2016.

[21] U.S. International Trade Commission, Digital Trade in the U.S. and Global Economies, Part 1, Publication No: 4415, Investigation No: 332-531, July 2013, p.6-1, https://www.usitc.gov/publications/332/pub4415.pdf.

[22] U.S. International Trade Commission, Digital Trade in the U.S. and Global Economies, Part 2, Publication No: 4485, Investigation No: 332-540, p.13, August 2014, https://www.usitc.gov/publications/332/pub4485.pdf.

[23] BCG, "Cross Border E-Commerce," September 18, 2014.

[24] Department of Commerce Economics and Statistics Administration, Digitally Deliverable Services Remain an Important Component of U.S. Trade, May 28, 2015, http://www.esa.doc.gov/economic-briefings/digitally-deliverableservices-remain-important-component-us-trade.

[25] U.S. International Trade Commission, Digital Trade in the U.S. and Global Economies, Part 2, Publication No: 4485, Investigation No: 332-540, August 2014, https://www.usitc.gov/publications/332/pub4485.pdf.

[26] James Manyika, Sree Ramaswamy, and Somesh Khanna, et al. Digital America: A Tale of the Haves and Have-Mores, McKinsey Global Institute, December 2015, p.40, http://www.mckinsey.com/industries/high-tech/ourinsights/digital-america-a-tale-of-the-haves-and-have-mores.

[27] Matthieu Pélissié du Rausas, James Manyika, and Eric Hazan, et al., Internet matters: The Net's sweeping impact on growth, jobs, and prosperity, McKinsey Global Institute, May 2011, p. 21, http://www.mckinsey.com/industries/hightech/our-insights/internet-matters.

[28] Google President Margo Georgiadis, Economic Impact United States 2014, p. 20, https://static.googleusercontent.com/media/www.google.com/en//economicimpact/reports/2014/ei-report-2014.pdf.

[29] Digital potential is defined as the upper bounds of digitization in the leading sectors included in the study. James Manyika, Sree Ramaswamy, and Somesh Khanna, et al. Digital America: A Tale of the Haves and Have-Mores, McKinsey Global Institute, December 2015, p. 32, http://www.mckinsey.com/industries/high-tech/our-insights/digitalamerica-a-tale-of-the-haves-and-have-mores.

[30] Ibid.

[31] Paul Zwillenberg, Dominic Field, and David Dean, Greasing the Wheels of the Internet Economy, Boston Consulting Group, February 2014. https://www.bcgperspectives.com/content/articles/digital_economy_telecommunications_greasing_wheels_internet_economy/.

[32] Ibid.

[33] The World Bank Group, World Development Report 2016: Digital Dividends, 2016, http://www.worldbank.org/en/publication/wdr2016.

[34] World Economic Forum; Global Competitiveness Report 2015-2016; Date of data collection or release: 1st September 2015; http://www.weforum.org/gcr.

[35] The World Bank Group, World Development Report 2016: Digital Dividends, 2016, http://www.worldbank.org/en/publication/wdr2016.

[36] http://www.state.gov/e/eb/cip/netfreedom/index.htm.

[37] Ambassador Michael B.G. Froman, "Getting Trade Right," Democracy Journal, Fall 2015, http://democracyjournal.org/magazine/38/getting-trade-right-1/. For more information, see also: The President of the United States, International Strategy for Cyberspace, May 2011, http://www.whitehouse.gov/sites/default/files/rss_viewer/international_strategy_for_cyberspace.pdf.

[38] John B. Morris, Jr., Twenty Years after the Birth of the Modern Internet, U.S. Policies Continue to Help the Internet Grow and Thrive, May 1, 2015, https://www.ntia.doc.gov/blog/2015/twenty-years-after-birth-modern-internet-uspolicies-continue-help-internet-grow-and-thriv.

[39] Alan B Davidson, "The Commerce Department's Digital Economy Agenda," November 9, 2015, https://www.commerce.gov/news/blog/2015/11/commerce-departments-digital-economy-agenda.

[40] Secretary of Commerce Penny Pritzker, "Commerce Launches Digital Attaché Program to Address Trade Barriers," March 11, 2016, https://www.commerce.gov/news/opinion-editorials/2016/03/commerce-launches-digital-attacheprogram-address-trade-barriers.

[41] Digitally intensive industries include sectors in communications, finance, trade, other services, and manufacturing. U.S. International Trade Commission, Digital Trade in the U.S. and Global Economies, Part 2, Publication No: 4485, Investigation No: 332-540, August 2014, pp. 106-108, https://www.usitc.gov/publications/332/pub4485.pdf.

[42] For more information on TPA, see CRS In Focus IF10038, Trade Promotion Authority (TPA), by Ian F. Fergusson, and CRS Report RL33743, Trade Promotion Authority (TPA) and the Role of Congress in Trade Policy, by Ian F. Fergusson, Trade Promotion Authority (TPA) and the Role of Congress in Trade Policy, by Ian F. Fergusson.

[43] OECD. (2015), OECD Digital Economy Outlook 2015, p. 38, OECD Publishing, Paris. DOI: http://dx.doi.org/10.1787/9789264232440-en.

[44] International Trade Administration, 2015 Top Markets Report Semiconductors and Semiconductor Manufacturing Equipment," July 2015, http://trade.gov/topmarkets/pdf/Semiconductors_Top_Markets_Report.pdf.

[45] Data on Harmonized System code 9018 from U.N. Comtrade: http://comtrade.un.org.

[46] CRS analysis of tariff data from the WTO Tariff Analysis Online (TAO): https://tao.wto.org.

[47] U.S. Census Bureau.

[48] Harmonized System code 8527, from WTO TAO.

[49] Trans-Pacific Partnership Annex 2-D: Vietnam Tariff Elimination Schedule, published by New Zealand Ministry of Foreign Affairs and Trade: https://www.mfat.govt.nz/assets/_securedfiles/trans-pacific-partnership/annexes/2-d.-vietnam-tariff-elimination-schedule.pdf.

[50] U.S. International Trade Commission, Digital Trade in the U.S. and Global Economies, Part 1, Publication No: 4415, Investigation No: 332-531, July 2013, p.5-1, https://www.usitc.gov/publications/332/pub4415.pdf.

[51] OECD. (2015), "Executive summary," OECD Digital Economy Outlook 2015, p. 5, OECD Publishing, Paris. DOI: http://dx.doi.org/10.1787/9789264232440-2-en.

[52] Rachael King, "AT&T to Move 80% of Its Applications to Cloud by Year's End," The Wall Street Journal, March 16, 2016, http://blogs.wsj.com/cio/2016/03/16/att-to-move-80-of-its-applications-to-cloud-by-years-end/.

[53] Google Cloud Platform Blog, "Google Cloud Platform adds two new regions, 10 more to come," March 22, 2016, https://cloudplatform.googleblog.com/2016/03/announcing-two-new-Cloud-Platform-Regions-and-10-more-tocome_22.html?mod=djemCIO_h.

[54] Steve Rosenbush, "Oracle's New Service Turns Cloud Computing 'Inside-Out'," The Wall Street Journal, March 24, 2016.

[55] Jay Greene, "Amazon to Launch Cloud Migration Service," The Wall Street Journal, March 15, 2016.

[56] Information Technology Industry Council, ITI Calls USTR Attention to Increasing use of Data Localization as a Trade Barrier and Threat to U.S. and Global Economic Growth, October 29, 2015, http://www.itic.org/news-events/news-releases/iti-calls-ustr-attention-to-increasing-use-of-data-localization-as-a-trade-barrier-and-threat-to-u-s-and-global-economic-growth.

[57] Galexia Consulting, 2016 BSA Global Cloud Computing Scorecard, Business Software Alliance, April 2016, http://cloudscorecard.bsa.org/2016/.

[58] Ambassador Michael B.G. Froman, 2016 National Trade Estimate Report on Foreign Trade Barriers, Office of the United States Trade Representative, 2016, https://ustr.gov/sites/default/files/2016-NTE-Report-FINAL.pdf.

[59] Intellectual property is a creation of the mind—such as an invention, literary/artistic work, design, symbol, name, or image—embodied in a physical or digital object.

[60] U.S. Department of Commerce, Intellectual Property and the U.S. Economy: Industries in Focus, prepared by the Economics and Statistics Administration and the U.S. Patent and Trademark Office, March 2012.

[61] Ibid.

[62] U.S. Bureau of Economic Analysis (BEA), U.S. Trade in Services data, released on October 15, 2015. The charges for the use of IP reflect those not included elsewhere in BEA services data.
[63] ITC, Digital Trade in the U.S. and Global Economies, Part 1, USITC Publication 4415, July 2013, p. 5-15.
[64] Envisional, Technical Report: An Estimate of Infringing Use of the Internet, January 2011.
[65] Frontier Economics, Estimating the Global Economic and Social Impacts of Counterfeiting and Piracy, report commissioned by Business Action to Stop Counterfeiting and Piracy (BASCAP), February 2011.
[66] USTR, 2015 Special 301 Report, April 2015, p. 13.
[67] OECD/EU Intellectual Property Office, Trade in Counterfeit and Pirated Goods: Mapping the Economic Impact, 2016. The study did not include online IP infringement, among other things.
[68] OECD, Enquiries into Intellectual Property's Economic Impact, August 10, 2015, p. 4.
[69] USTR, 2015 Out-of-Cycle Review of Notorious Markets, December 2015, p. 9.
[70] ITC, Digital Trade in the U.S. and Global Economies, Part 1, USITC Publication 4415, July 2013, p. 5-1.
[71] Ibid., p. 5-15.
[72] Computer & Communications Industry Association (CCIA), Comments to USTR in Response to Request for Public Comments to Compile the National Trade Estimate Report on Foreign Trade Barrier, 2015.
[73] Ibid.
[74] For more information on fair use, please see CRS Report RS22801, General Overview of U.S. Copyright Law, by Brian T. Yeh.
[75] Ibid.
[76] USTR, 2016 National Trade Estimate Report on Foreign Trade Barriers, p. 179, March 2016.
[77] Eva Dou, "China's Great Firewall Gets Taller," The Wall Street Journal, January 30, 2015.
[78] "Beijing is banning all foreign media from publishing online in China," Quartz, February 18, 2016.
[79] Max Seddon, "Russia's chief internet censor enlists China's know-how," Financial Times, April 26, 2016.
[80] Mark Scott, "Google Fined by French Privacy Regulator," The New York Times, March 24, 2016.
[81] Alan McQuinn, "France Demands Right to Censor the Global Internet," The Innovation Files, March 28, 2016.
[82] For more information on FCC rules on net neutrality, see CRS Report R43971, Net Neutrality: Selected Legal Issues Raised by the FCC's 2015 Open Internet Order, by Kathleen Ann Ruane, and CRS Report R40616, Access to Broadband Networks: The Net Neutrality Debate, by Angele A. Gilroy.
[83] Julia Fioretti. "EU regulators take tough approach to net neutrality," Reuters, June 2, 2016. Note: under a zero rating, a provider can exempt traffic from certain sites and services from a user's monthly data allowance.
[84] For more information on cybersecurity, see CRS Report R43831, Cybersecurity Issues and Challenges: In Brief, by Eric A. Fischer.
[85] Joseph Marks, "Indictment: Iranians made 'coordinated' cyberattacks on U.S. banks, dam," Politico Pro, March 24, 2016. For more information, see CRS Insight IN10259, Attribution in Cyberspace: Challenges for U.S. Law Enforcement, by Kristin Finklea.

[86] For more information on encryption, see CRS Report R44187, Encryption and Evolving Technology: Implications for U.S. Law Enforcement Investigations, by Kristin Finklea, and CRS Report R44407, Encryption: Selected Legal Issues, by Richard M. Thompson II and Chris Jaikaran.
[87] Based on Bureau of Economic Analysis (BEA), World Bank, and United Nations Committee on Trade and Development data.
[88] Joshua P. Meltzer, The Importance of the Internet and Transatlantic Data Flows for U.S. and EU Trade and Investment , Brookings, p.12, October 2014, http://www.brookings.edu/research/papers/2014/10/internet-transatlanticdata-flows-meltzer.
[89] Ibid, p. 17.
[90] All figures on U.S.-EU trade and data flows includes the United Kingdom (UK) as part of the EU. Without the UK, the statistics would be lower.
[91] European Commission Press Release, "European Commission launches EU-U.S. Privacy Shield: stronger protection for transatlantic data flows," July 12, 2016, http://europa.eu/rapid/press-release_IP-16-2461_en.htm.
[92] For more information on the Safe Harbor Agreement or Privacy Shield, please see CRS Report R44257, U.S.-EU Data Privacy: From Safe Harbor to Privacy Shield, by Martin A. Weiss and Kristin Archick.
[93] Daniel R. Stoller, "EU-U.S. Data Transfer Privacy Shield Opinion Imminent," Bloomberg BNA, April 5, 2016.
[94] USTR, 2016 National Trade Estimate Report on Foreign Trade Barriers, p. 178.
[95] European Commission, "Commission updates EU audiovisual rules and presents targeted approach to online platforms," Press Release, May 25, 2016.
[96] European Commission, "Agreement on Commission's EU data protection reform will boost Digital Single Market," Press Release, December 15, 2015.
[97] Ali Qassim, "Lack of EU Data Reg Guidance Has Companies Uncertain," Bloomberg BNA, April 26, 2016.
[98] Stephen Gardner, "Effective Date Set for EU General Data Protection Rule," Bloomberg BNA, May 4, 2016.
[99] Ali Qassim, "U.K. Privacy Office Seeks EU-Compliant Laws Despite Brexit," Bloomberg BNA, July 5, 2016.
[100] E-Marketer, Ecommerce Drives Retail Sales Growth in China, September 25, 2015.
[101] The text of the agreement is available at: https://www.whitehouse.gov/the-press-office/2015/09/25/fact-sheet-uschina-economic-relations.
[102] For more information on U.S. China trade relations and the BIT negotiations, see CRS In Focus IF10030, U.S.- China Trade Issues, by Wayne M. Morrison, and CRS In Focus IF10307, A U.S.-China Bilateral Investment Treaty (BIT): Issues and Implications, by Wayne M. Morrison.
[103] The White House, "FACT SHEET: President Xi Jinping's State Visit to the United States," September 25, 2015.
[104] U.S. agencies included the Departments of Justice, Homeland Security, and State and the National Security Council and Intelligence Community and while Chinese representatives came from the Committee of Political and Legal Affairs of CPC Central Committee, Ministry of Public Security, Ministry of Foreign Affairs, Ministry of Industry and Information Technology, Ministry of State Security, Ministry of Justice and the State Internet Information Office. Department of Justice, "First U.S.-China High-Level Joint Dialogue on Cybercrime and Related Issues Summary of Outcomes," December 2, 2015.

[105] Department of Homeland Security, "Second U.S.-China Cybercrime and Related Issues High Level Joint Dialogue," June 15, 2016.

[106] United National Conference on Trade and Development, Data protection regulations and international data flows: Implications for trade and development, 2016, http://unctad.org/en/PublicationsLibrary/dtlstict2016d1_en.pdf.

[107] Ibid, p.56.

[108] Ibid, p. xi.

[109] Koske,I., et al. (2014), "The Internet Economy - Regulatory Challenges and Practices", OECD Economics Department Working Papers, No. 1171, OECD Publishing, Paris. DOI: http://dx.doi.org/10.1787/5jxszm7x2qmr-en.

[110] For more information, see https://www.wto.org/english/tratop_e/serv_e/serv_e.htm and CRS Report R43291, U.S. Trade in Services: Trends and Policy Issues, by Rachel F. Fefer.

[111] For more information on WTO and the Doha Round, see CRS In Focus IF10002, The World Trade Organization, by Ian F. Fergusson and Rachel F. Fefer, and CRS Insight IN10422, The WTO Nairobi Ministerial, by Rachel F. Fefer.

[112] World Trade Organization, WTO members conclude landmark $1.3 trillion IT trade deal, December 16, 2015, https://www.wto.org/english/news_e/news15_e/ita_16dec15_e.htm.

[113] Office of the U.S. Trade Representative, U.S. and WTO Partners Announce Final Agreement on Landmark Expansion of Information Technology Agreement, December 2015, https://ustr.gov/about-us/policy-offices/pressoffice/press-releases/2015/december/US-WTO-Partners-Announce-Final-Agreement-on-Expansion-ITA.

[114] "Exclusively for the purposes of the work programme, and without prejudice to its outcome, the term 'electronic commerce' is understood to mean the production, distribution, marketing, sale or delivery of goods and services by electronic means."

[115] For more information, see https://www.wto.org/english/tratop_e/ecom_e/ecom_ briefnote _e.htm.

[116] For more information on the Work Programme on Electronic Commerce, see https://www.wto.org/english/tratop_e/ecom_e/ecom_e.htm and https://www.wto.org/english/thewto_e/minist_e/min99_e/english/about_e/20ecom_e.htm.

[117] https://www.wto.org/english/news_e/news15_e/gc_30nov15_e.htm.

[118] For pharmaceutical products, the implementation period has been extended until January 1, 2033.

[119] WTO, "Overview: The TRIPS Agreement," https://www.wto.org/english/tratop_e/trips_e/intel2_e.htm. For more information, see CRS Report RL34292, Intellectual Property Rights and International Trade, by Shayerah Ilias Akhtar and Ian F. Fergusson.

[120] WTO, General Council, "Item 6 – Work Programme on Electronic Commerce – Review of Progress," WT/GC/W/701, July 24, 2015.

[121] BSA, Powering the Digital Economy: A Trade Agenda to Drive Growth; and BSA, Shadow Market: 2011 BSA Global Software Piracy Study, May 2012.

[122] U.S. Congress, Senate Committee on Foreign Relations, WIPO Copyright Treaty (WCT) (1996) and WIPO Performances and Phonograms Treaty (1996), Report to accompany treaty document 105-17, 105th Cong., 2nd sess., October 14, 1998, S.Exec. Rept. 105-25.

[123] See P.L. 105-304.

[124] For more information on this statute, see CRS Report R43436, Safe Harbor for Online Service Providers Under Section 512(c) of the Digital Millennium Copyright Act, by Brian T. Yeh.

[125] USTR, 2016 Special 301 Report, April 2016.

[126] Joshua Paul Meltzer, Maximizing the Opportunities of the Internet for International Trade, The E15 Initiative, January 2016, http://e15initiative.org/publications/maximizing-opportunities-internet-international-trade/.
[127] Michael Froman, "We are at the end of the line on the Doha Round of trade talks," Financial Times, December 13, 2015.
[128] WTO, "Non-Paper from the United States," JOB/GC/94, July 4, 2016.
[129] Ibid.
[130] https://ustr.gov/sites/default/files/uploads/agreements/fta/singapore/asset_upload_file708_4036.pdf.
[131] This statement was used in U.S. free trade agreements with Australia, Bahrain, Colombia, Central America and the Dominican Republic, Morocco, Oman, Panama, Peru, and South Korea. Chile used a slightly different text.
[132] For more information on KORUS, see CRS Report RL34330, The U.S.-South Korea Free Trade Agreement (KORUS FTA): Provisions and Implementation, coordinated by Brock R. Williams.
[133] KORUS FTA, Chapter 13, Annex 13-B, Section B. https://ustr.gov/sites/default/files/uploads/agreements/fta/korus/asset_upload_file35_12712.pdf.
[134] For more on TPP, see CRS In Focus IF10000, TPP: An Overview, by Brock R. Williams and Ian F. Fergusson, CRS Report R44489, The Trans-Pacific Partnership (TPP): Key Provisions and Issues for Congress, coordinated by Ian F. Fergusson and Brock R. Williams, and CRS In Focus IF10390, TPP: Digital Trade Provisions, by Rachel F. Fefer.
[135] Industry Trade Advisory Committee on Information and Communication Technologies, Services, and Electronic Commerce (ITAC 8), Advisory Committee Report to the President, the Congress and the USTR on the TPP Trade Agreement, December 3, 2015, https://ustr.gov/sites/default/files/ITAC-8-Information-and-CommunicationTechnologies-Services-and-Electronic-Commerce.pdf.
[136] Annex I for Japan includes Foreign Exchange and Foreign Trade Law (Law No. 228 of 1949), Article 274 Cabinet Order on Foreign Direct Investment (Cabinet Order No. 261 of 1980), Article 3, https://ustr.gov/sites/default/files/TPPFinal-Text-Annex-I-Non-Confor ming-Measures-Japan.pdf.
[137] TPP Chapter 14, Article 14.8.2.
[138] For more, see CRS In Focus IF10390, TPP: Digital Trade Provisions, by Rachel F. Fefer.
[139] For more on TiSA, see CRS In Focus IF10311, Trade in Services Agreement (TiSA) Negotiations, by Rachel F. Fefer, and CRS Report R44354, Trade in Services Agreement (TiSA) Negotiations: Overview and Issues for Congress, by Rachel F. Fefer.
[140] Inside U.S. Trade, "Despite 'TISA-Plus' Aims, EU's E-Commerce Proposal For T-TIP Falls Short", August 13, 2015.
[141] Ibid.
[142] European Parliament Resolution of 8 July 2015 Containing the European Parliament's Recommendations to the European Commission on the Negotiations for the Transatlantic Trade and Investment Partnership, July 8, 2015.
[143] See, for example, Trans-Atlantic Business Council (TABC), TABC Position Statement on Cross-Border Data flows, May 30, 2014.
[144] See, for example, Transatlantic Consumer Dialogue (TACD), "Resolution Data Flows in the TransAtlantic Trade and Investment Partnership," October 2013.
[145] USTR, "U.S. Objectives, U.S. Benefits in the Transatlantic Trade and Investment Partnership: A Detailed View," fact sheet, March 2014.

[146] David Ohrenstein, "Eight Ways the TTIP Can Be a Global Blueprint for Digital Trade," TechPost, May 15, 2013.
[147] The Group of Twenty (G-20) is a forum for advancing international cooperation and coordination among 20 major advanced and emerging-market economies. The G-20 includes Argentina, Australia, Brazil, Canada, China, France, Germany, India, Indonesia, Italy, Japan, Mexico, Russia, Saudi Arabia, South Africa, South Korea, Turkey, United Kingdom, and the United States, as well as the European Union (EU). For more information on the G-20, see CRS Report R40977, The G-20 and International Economic Cooperation: Background and Implications for Congress, by Rebecca M. Nelson.
[148] http://g20.org.tr/g20-leaders-commenced-the-antalya-summit/.
[149] Joint Declaration by G7 ICT Ministers, April 30, 2016, http://www.g8.utoronto.ca/ict/2016-ict-declaration.html.
[150] http://www.oecd.org/internet/ministerial/. The G-7 is a subset of the G-20 and includes: Canada, France, Germany, Italy, Japan, United Kingdom and the United States.
[151] OECD Ministerial Declaration, May 2016, http://www.oecd.org/sti/ieconomy/Digital-Economy-MinisterialDeclaration-2016.pdf.
[152] Asia Pacific Economic Cooperation (APEC) is a regional economic forum established in 1989 with 21 Asian Pacific economies as members. http://www.apec.org/About-Us/About-APEC.aspx.
[153] http://www.apec.org/Groups/Committee-on-Trade-and-Investment/Electronic-Commerce-Steering-Group.aspx.

In: U.S. Technological Endeavors
Editor: Beverly Howard
ISBN: 978-1-53610-547-6
© 2017 Nova Science Publishers, Inc.

Chapter 2

DIGITAL ECONOMY AND CROSS-BORDER TRADE: THE VALUE OF DIGITALLY-DELIVERABLE SERVICES[*]

Jessica R. Nicholson and Ryan Noonan

EXECUTIVE SUMMARY

The digital, or Internet, economy has transformed many aspects of our lives over the past two decades. How we communicate, entertain ourselves, make decisions, and do business continues to evolve as the digital economy grows in size and importance. Given this transformation, it is becoming even more important for policymakers to consider how the Internet affects our lives and the economy as a whole.

The Department of Commerce has played an instrumental role in developing policies that facilitate the digital economy. The Department's Internet Policy Task Force identifies leading public policy and operational challenges in the Internet environment. The Task Force is committed to maintaining the global free flow of information online. This report provides a framework for understanding the size and nature of some cross border data flows. We do not forecast how data flows would change in response to any given policy decision.

---

[*] This is an edited, reformatted and augmented version of ESA Issue Brief # 01-14, issued by the Economics and Statistics Administration, U.S. Department of Commerce, January 27, 2014.

Our analysis uses digitally-enabled services categories identified in previous research as a starting point for identifying "digitally-deliverable" services—i.e., services that may be, but are not necessarily, delivered digitally. These service categories are the ones in which digital technologies present the greatest opportunity to transform the relationship between buyer and seller from the traditional in-person delivery mode to a digital one. The aggregate trade data used in this report capture a mix of transactions that are entirely digital, somewhat digital, or entirely non-digital; our estimates focus on those transactions that are most likely to be done digitally. Because we do not have a direct measure of services that are digitally traded within each service category, it is difficult to precisely estimate how much this percentage has increased over time. Instead, this paper presents an upper-bound estimate of the percentage of services exports that are digitally-deliverable.

We also know that there is a great deal of interest in understanding the economic value related to digitally-enabled data and services delivered to users at no price. An exact number of bytes associated with these data flows may not be known, but the number is large and growing. Data flows related to subscribers' use of Facebook, Google, Twitter, and other free online services are not covered in this report because they are not captured as monetary transactions in Federal cross-border trade statistics. However, traditional services trade related to the operation of these free services—such as advertising services, Internet access services, and legal services—are included in our analysis if they are exported to other countries from the United States.

Our analysis reveals that in 2011:

- The United States exported $357.4 billion in digitally-deliverable services. This represented over 60 percent of U.S. services exports and about 17 percent of total U.S. goods and services exports.1
- The United States imported $221.9 billion in digitally-deliverable services. This represented 56 percent of U.S. services imports and about 8 percent of total U.S. goods and services imports.
- The United States had a digitally-deliverable services trade surplus of $135.5 billion.
- The total value of digitally-deliverable services in the supply chain of total U.S. goods and services exports was $627.8 billion, or about 34 percent of total export value.
- The majority of U.S. digitally-deliverable services exports went to Europe and to the Asia and Pacific region.

- Specifically, the United States exported the highest value of digitally-deliverable services to the United Kingdom, Canada, Ireland, and Japan. The highest values of digitally-deliverable imports came from the United Kingdom, Bermuda, Switzerland, and Canada.

INTRODUCTION

The digital, or Internet, economy has transformed many aspects of our lives the past two decades. How we communicate, entertain ourselves, make decisions, and do business continues to evolve as the digital economy grows in size and importance. Given this transformation, it is becoming more important for policymakers to consider how the Internet affects our lives and the economy as a whole.

The Department of Commerce has played an instrumental role in developing policies that facilitate the digital economy. The Department's Internet Policy Task Force identifies leading public policy and operational challenges in the Internet environment. The Task Force is committed to maintaining the global free flow of information online. This report provides a framework for understanding the size and nature of some cross border data flows. We do not forecast how data flows would change in response to any given policy decision.

A portion of cross-border data flows are associated with transactions that are captured in traditional services trade statistics. For example, an engineering firm or a financial services firm exporting their services likely transmits data across the U.S. border in the course of day-to-day-activities. While the number of bytes of data associated with these cross-border transactions is not measured like tons of coal at the border, the dollar value associated with cross-border services transactions is collected and published by the Bureau of Economic Analysis (BEA).

This paper estimates the economic value associated with cross-border data flows by first determining which services trade categories most likely represent transactions that take place online rather than by other means, such as people crossing a border to perform the service. Previous research has identified a number of service categories as digitally-enabled—i.e., those services for which information and communications technologies (ICT) demonstrably play an important role in facilitating cross-border trade.

Services that are not primarily delivered online are excluded from our analysis. For example, education is a category for which the primary mode of delivery is in-person rather than digital and is therefore not included in digital trade. An exception is distance learning, which is primarily delivered online. Distance learning is not part of the education services trade category and is captured in official statistics as training, a component of business, professional, and technical services.

It is important to note that these digitally-deliverable services are not new trade categories. Rather, estimates of digitally-deliverable services trade are based on an approximation of *how* trade in an existing category of services likely took place. The aggregate trade data capture a mix of transactions that are entirely digital, somewhat digital, or entirely non-digital; our estimates focus on those transactions that are most likely to be done digitally. The degree of "digital-ness" of transactions in the various services categories is also likely changing over time as businesses respond to changes in their markets and to new technology.

Because we do not have a direct measure of services that are digitally traded within each service category, it is difficult to precisely estimate how much this percentage has increased over time. Instead, this paper presents an upper-bound estimate of the percentage of services exports that are digitally-deliverable.

WHAT ARE CROSS-BORDER DATA FLOWS?

Data cross country borders in a variety of ways. They are constantly flowing from the United States to other countries, and vice versa. Data flows can be measured in bytes of digital traffic per second, hour, year, or any other amount of time. Data flows can also be categorized in different ways—for example, by the commercial characteristics of the contents. Below, we define four categories of data flows:

1. Purely non-commercial traffic, such as government or military communications.
2. Commercial data and services exchanged between businesses or other related-parties, such as supply chain information, personnel data, or design information, at a $0 market price.
3. Flows of data that are traded between a seller and buyer at a market price. This may include royalty payments associated with movies, TV, or music sales; online banking services; advertising; or other transactions.

4. Digitally-enabled data and services delivered to or from end-users at a $0 market price. Examples include free email services, search engine services, map and direction services, and social media services.

For the purpose of this report, we focus on providing an estimate of the potential size of the third category, digital trade in services. Federal economic statistics collected by the Bureau of Economic Analysis (BEA) provide estimates of the dollar values associated with trade in services, but BEA data are unable to quantify the size of information flows in terms of bytes. The remaining three categories require a non-financial understanding of data flows, and Federal economic statistics do not provide a sound basis for estimating the size of the flows associated with those transactions.

We know that there is a great deal of interest in understanding the economic value related to the fourth category above – digitally-enabled data and services delivered at no price. An exact number of bytes associated with these data flows may not be known, but the number is large and growing. Data flows related to subscribers' use of Facebook, Google, Twitter, and other free online services are not covered in this report because they are not captured as monetary transactions in Federal cross-border trade statistics. However, traditional services trade related to the operation of these free services—such as advertising services, Internet access services, and legal services—are included in our analysis if they are exported to other countries from the United States.

Finally, there is one caution to keep in mind. The default way of thinking about digital trade in books, movies or music tends to be in terms of transactions that an individual might make in purchasing an e-book, movie, or sound recording from an online store or in subscribing to a service that provides on-demand access to a catalog of movies or music. However, those types of transactions do not typically take place across borders. In fact, in most cases, they are specifically blocked because of the geographically-specific legal regimes that govern intellectual property associated with such content. The correct way of thinking about the dollars exchanged in these categories is, for example, the exchange of money from a firm in one country to a firm in another country for the right to sell content that is protected by intellectual property laws in a given geographical area.[2]

If the copyrighted works were supplied to foreign consumers from a U.S. business through a U.S. affiliate located in the country of the purchaser, the transaction would be captured by BEA as services supplied through affiliates rather than as cross-border trade.[3]

DIGITALLY-DELIVERABLE SERVICES DEFINED

To understand how the digital economy contributes to cross-border trade in services, one approach is to identify or estimate those services that are delivered to end-users digitally. In a recent study, BEA uses the term "digitally-enabled services" to refer to those services that are principally or largely enabled by information and communication technologies (ICT).[4,5] To measure the size of digitally-enabled trade, BEA follows the definition used by the United Nations Conference on Trade and Development (UNCTAD). In 2007, UNCTAD defined seven categories in the International Monetary Fund balance of payments accounts as ICT-enabled: communications services; insurance; financial services; computer and information services; royalties and license fees; other business services; and personal, cultural, and recreational services.

These categories correspond to five categories in BEA's cross-border trade statistics:

- Business, professional, and technical services (except construction);
- Royalties and license fees;
- Insurance services;
- Financial services; and
- Telecommunications.

It is important to note that there is no way to determine an exact percentage of the trade in each of these categories that was actually delivered digitally, as the international trade statistics do not capture how services are provided. There are also other types of services that may be traded digitally, but the UN and BEA have chosen to exclude these services from analysis because their primary mode of delivery is not digital. We discuss these limitations in the Technical Appendix of this report.

This report uses the term "digitally-deliverable" rather than "digitally-enabled" to describe these five service categories. Because there are no data available that indicate whether these services were actually delivered digitally or by some other means, the term "digitally-deliverable" is intended to convey that these services *may* be delivered digitally. In particular, these service categories are the ones in which digital technologies present the most opportunity to transform the relationship between buyer and seller from the traditional in-person delivery mode to a digital one.

Which Services are Included?

The cross-border trade in services categories discussed in this report and defined as digitally-deliverable include a wide array of services. To make these concepts more concrete, here are some examples of the kinds of services included in these categories.

Business, professional, and technical services: The largest category of digitally-deliverable services includes a large number of activities. Some of these activities, such as computer and information services, are inherently digital, although some computer products provided on a physical medium are classified as goods. Others, such as legal services or architectural services, are known to make intensive use of digital resources. Legal briefs and documents, consulting reports, and architectural or engineering designs can easily be digitized and transmitted over the Internet to customers located anywhere in the world. Advertising services are used throughout the digital world, whether by department stores, sports teams, or high-tech companies like Facebook and Google.

Royalties and license fees: This category covers charges for the use of intellectual property, such as patents, trademarks, copyrights, industrial processes and designs, and franchises. Two-thirds of royalties and license fees are for industrial processes and for general use computer software. Improvements in digital communications have enabled easier sharing of schematics and designs across borders. The value associated with reproducing and disseminating copyrighted materials in digital form (including e-books, music, movies, and software) is included here if a U.S. company directly supplies the content to a foreign consumer or business, or vice versa.

Financial services: Banking is an increasingly digital activity. Whether customers are accessing information, managing their accounts, paying bills, or transferring funds, they are doing it online far more often. Investment activities, whether they are market research, financial planning tools, or buying and selling shares in the stock market or a mutual fund, are also increasingly being performed online and across borders.

Insurance services: Insurers are increasingly turning to digital networks for communication and the delivery of products. Digitized paper documents and e-delivery are growing in importance throughout the industry. As with banking services, customers are turning more and more to the digital transmission of premiums, and they are receiving payments for claims in the same manner. In addition, reinsurance—the purchase of insurance policies by another insurance company as a strategy for mitigating risk—involves transfers of funds and documents that may be conducted digitally.

Telecommunications: The telecommunications service category is arguably the most digital of the service categories studied in this report. Telephone calls, video conferences, e-mail, and voice mail are highly digitized activities. Internet access services are also included in this category.

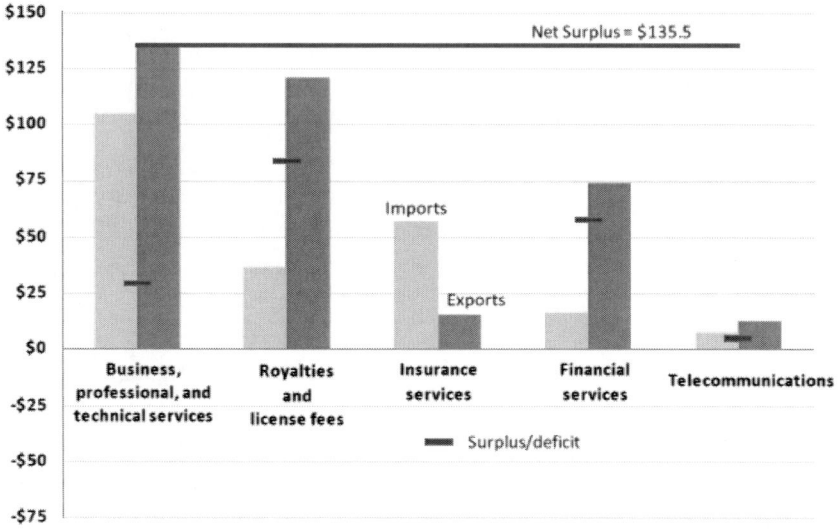

Source: Economics and Statistics Administration analysis using data from the Bureau of Economic Analysis.

Figure 1. U.S. Trade in Digitally-Deliverable Services, 2011. (billions of dollars)

UNITED STATES CROSS-BORDER TRADE DATA

Using these five service categories, we find that digitally-deliverable services accounted for over 60 percent of U.S. service exports in 2011, and about 17 percent of overall U.S. goods and services exports.[6] In dollar terms, U.S. exports of digitally-deliverable services totaled $357.4 billion, while imports totaled $221.9 billion, resulting in a trade surplus of $135.5 billion.[7]

Business, professional, and technical services accounted for the largest share of digitally-deliverable services exports (38 percent), followed by royalties and license fees (34 percent), financial services (21 percent), insurance services (4 percent), and telecommunications (4 percent) (see Figure 1). In 2011, 47 percent of digitally-deliverable services imports were business, professional, and technical services, followed by insurance services (26 percent), royalties and license fees (17 percent), financial services (7 percent), and telecommunications (3 percent).

Overall, the United States saw a $135.5 billion digitally-deliverable trade surplus in 2011. The only individual category that registered a deficit was insurance services.

Data Flows and Trade Policy

The Internet is a key platform for commerce between buyers and sellers located in different countries. As the Internet enables cross-border data flows, it also underpins much of global economic integration and international trade. For instance, cross-border data flows are now very common in business, from Internet-based communications like email and platforms such as eBay and Facebook that bring buyers and sellers together, to financial transactions to complete product purchases across borders, to the downloading of digital products and services.

Major concerns for U.S. policymakers regarding the digital economy include protecting privacy and security (i.e., protection of personal and business data), promoting the free flow of information, and avoiding localization requirements. Strong privacy and security safeguards protect the integrity of financial transactions and reassure consumers that their personal information will not be misused if they participate in digital commerce. Furthermore, well-conceived online security measures can help businesses protect the privacy of their employees, safeguard their assets, and maintain their competitive edge. In short, online commerce can thrive only if all users, whether buyers or sellers and regardless of nationality, trust that the system provides robust security and privacy protections.

Conversely, Internet-based trade is hindered by restrictions on data flows, such as limitations on data flows across national borders or other artificial restrictions. While some countries cite important privacy concerns as a basis for restricting data flows, such restrictions may be economically harmful and damage the free exchange of ideas, goods, and services online. While balanced and clear intellectual property policies promote the free flow of information as well as the rights of innovators, restrictions on the flow of information may be implemented in an unfair or nontransparent manner or may be more expansive than necessary.

> While localization requirements may serve some national goals, they can serve as disguised trade barriers when they unreasonably differentiate between domestic and foreign products or services. For example, as technologies increasingly rely on remote servers for data storage, access, and transfer, requirements that data must reside on local servers can increase costs by reducing economies of scale. Policies that discriminate against foreign technologies or require use of country-specific technology standards may reduce data security, interoperability, and the stability of the digital service. Localization requirements carry great risk of limiting the Internet's global character, making cross-border trade difficult for large companies and practically impossible for small businesses that cannot afford to implement separate systems and standards in every country in which they do business.

A NEW UNDERSTANDING OF TRADE IN SERVICES

Trade in all services accounts for nearly 30 percent of U.S. exports. On its own, this is a fairly large number, contributing over $600 billion to U.S. gross domestic product in 2011. One thing this simple valuation overlooks, however, is the value of services as inputs throughout the production process. Virtually any good that is exported requires services somewhere in its supply chain, whether those services are legal or business services, royalties or license fees, or even education or health care for production workers. Because such services are intermediate inputs, their value is absorbed into the final value of the good that is produced after the services are consumed. Simply adding up the services exported by the U.S. underestimates the true value of services to the U.S. export economy.

One way to estimate the true value of services to U.S. exports is by using input-output (I-O) tables, which show the interdependencies between industries and commodities throughout the economy and the production process. Using I-O tables, the Organisation for Economic Cooperation and Development (OECD) and the World Trade Organization (WTO) have conducted a study of Trade in Value Added (TiVA). According to their study, services as intermediate inputs accounted for nearly 50 percent of the value of exports from the U.S. in 2009, a number that is substantially higher than the 30 percent accounted for by final service exports.[8]

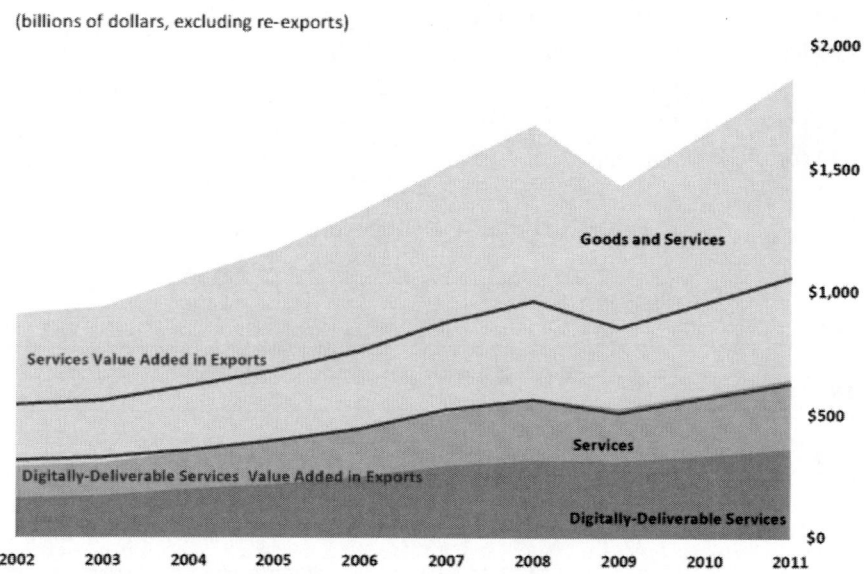

Source: Economics and Statistics Administration analysis using data from the Bureau of Economic Analysis.

Figure 2. U.S. Exports, 2002-2011.

To put this number into context, while final service exports are valued at over $600 billion, services may account for $1.1 *trillion* of export value throughout the entire production process.

To check the OECD and WTO numbers, we conducted a similar analysis using the I-O tables produced by BEA. In the end, we found that services as intermediate inputs accounted for a similar but slightly larger amount—57 percent— of the value of exports in 2011 (see Figure 2).

The Value of Digitally-Deliverable Services as Intermediate Inputs

The TiVA project makes clear that simply measuring trade in services does not necessarily provide a clear picture of the value of services as intermediate inputs in the production of goods or even other services. Combining the methodology used by the TiVA project with our categories of digitally-deliverable services, we can take our understanding of digitally-deliverable services one step further. BEA's I-O tables can be used to calculate the percentage of export value contributed by these services throughout the production process. The I-O tables use a different classification system than the cross-border trade in services data. Please refer to the Technical Appendix for the detailed industry concordance and a description of the methodology used for this calculation.

Table 1. Digitally-Deliverable Services

Year	As a Percentage of Products in Exports†	As a Percentage of Value Added in Exports
2002	17.6%	34.5%
2003	18.0%	34.6%
2004	18.6%	33.5%
2005	18.5%	33.7%
2006	18.5%	33.4%
2007	19.4%	34.5%
2008	18.8%	33.5%
2009	21.8%	35.5%
2010	20.3%	34.5%
2011	19.5%	33.6%

Source: Economics and Statistics Administration analysis using data from the Bureau of Economic Analysis
†For consistency with the input-output tables, these percentages exclude re-exports.

Our findings are displayed in Table 1 and Figure 2.[9] Recall that digitally-deliverable services account for about one-fifth of total U.S. exports When considered as inputs in the production process, we find that the contribution of digitally-deliverable services jumps to more than one-third of the total value of U.S. exports.

Because we are unable to determine exactly what portion of output in each of these industries is actually delivered digitally, it is difficult to draw more specific conclusions about trends over time. For instance, the production and delivery processes in the newspaper publishing industry may be substantially or primarily digital today but were less so a decade ago. Our analysis, however, cannot distinguish between the newspaper industry of today and the newspaper industry of ten years ago.

To provide some context about the increasing digitization of the industries in our study, we looked at data on revenues in some of the service industries and trends in the percentage of revenues attributed to e-commerce (see Figure 3). E-commerce sales are not a comprehensive measure of the digitization of these industries. Some services may not be purchased digitally, but may be delivered digitally, and therefore, would not necessarily be captured here. 10 Data on e-commerce revenues, although limited, do support the claim that these services are becoming increasingly digitized.[11]

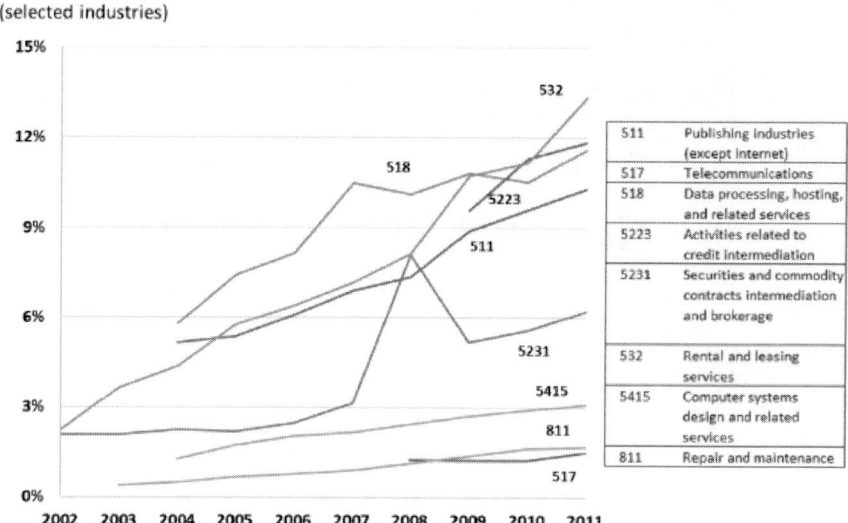

Source: Econoics and Statisics administration analysis using data from the U.S. Census bureau.
Note: Data are not available for all industries for all years.

Figure 3. Percentage of Revenues Contributd by E-Commerce, 2002-2011.

All of the industries for which data are available show positive trends in revenues attributed to e-commerce. These data are not conclusive evidence that all of the digitally-deliverable service industries included in our study are becoming increasingly digitally-delivered, nor are they a precise measure of the level of digitization in these service industries; however, these data do support the notion that the digital delivery of services is growing.

Even though data on the digitization of service exports are limited, there is no shortage of anecdotal evidence that an increased use of digitization is occurring in many of these industries. Media companies provide one example of this change. Newspapers (product of I-O industry 51110), book stores (product of I-O industry 511130), and music stores (product of I-O industry 512200) continue to shut down, while revenues from online media and entertainment continue to grow.

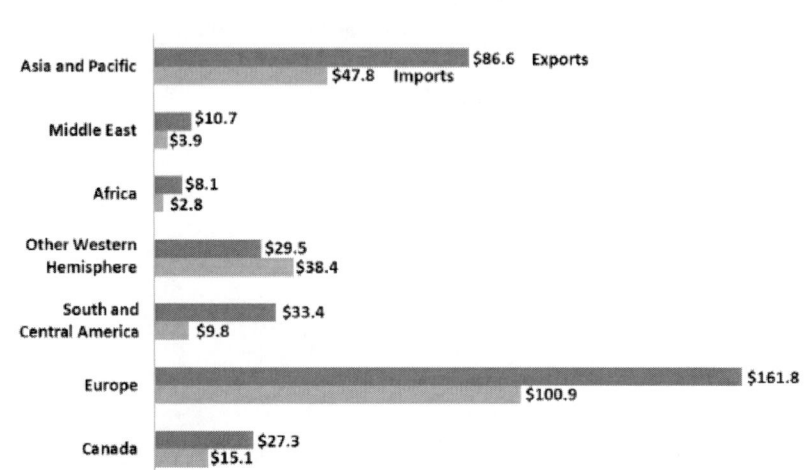

Source: Economics and Statistics Administration analysis using data from the Bureau of Economic Analysis (BEA).
Note: Region definitions can be found on the BEAwebsite at:
http://www.bea.gov/international/bp_web/geographic_area_definitions.cfm.

Figure 4. U.S.Exports and impors of Digitally-Deliverable Services, by region or country of destination or origin, 2011.

PricewaterhouseCoopers estimates that revenues for these online services will increase by approximately 13 percent a year for the next five years. IFPI, a trade group, reports that global music industry revenues-including both digital and non-digital revenues—grew by 0.3 percent in 2012—the best result since 1998—due to the growth of digital music including download sales, subscription services, music video streaming, digital radio, performance rights, and sychronization.[12]

Advertising and related services (I-O industry 541800) provides another example of increased digitization. According to *The Economist*, firms that depended on print advertising have been the worst hit by digitization of these services, while advertising on the web grows in leaps and bounds.13 In 2012, online ads were worth $88 billion, or 18 percent of global advertising spending, up from 2006, when online ads composed just 7 percent of global spending.14 This share is expected to continue growing.

CROSS-BORDER TRADE BY REGION AND SERVICE CATEGORY

To gain additional insight into how these service categories fit into our overall economy and trade policy, we can look at U.S. exports and imports by trading partner and by service category. To do this, we turn back to the BEA cross-border trade data.

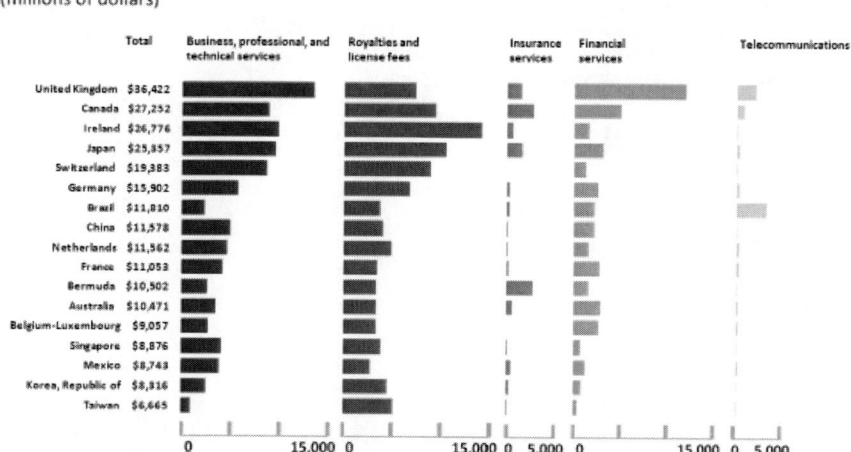

Source: Economics and Statistics Admiistraion analysis using data from the bureau of Economic Analysis.

Note: Data on export of insurance services to Belgium-Luxemboug and Switzerland are suppressed to avoid discosures if data individual conmanies.

Figure 5. U.S. Exports of Digitally-DeliverableServices by Service Category and Selected Econoics of Origin, 2011.

Regionally, the United States trades the greatest value of digitally-deliverable services with countries in Europe and the Asia and Pacific region (See Figure 4).[15,16] In 2011, trade with Europe accounted for 45 percent of the total value of both digitally-deliverable service exports and imports. This is down slightly from 2006 when this region accounted for 50 percent of exports and 51 percent of imports within these service categories.

The overall proportion of U.S. digitally-deliverable services exports and imports going to and coming from the Asia and Pacific region has not significantly changed over the past five years—exports to this region accounted for 23 percent of all exports in this category in 2006 compared to 24 percent in 2011, while imports accounted for 20 percent of the U.S. total in 2006 compared to 22 percent in 2011. However, within this region, the countries with which the U.S. conducts digital trade have changed. While Japan remains a top U.S. trading partner in the region, trade of digitally-deliverable services has increased with China and India. Trade of digitally-deliverable services with countries in South and Central America accounted for 9 percent of all U.S. digitally-enabled services exports and 4 percent of U.S. digitally-deliverable services imports in 2011.[17]

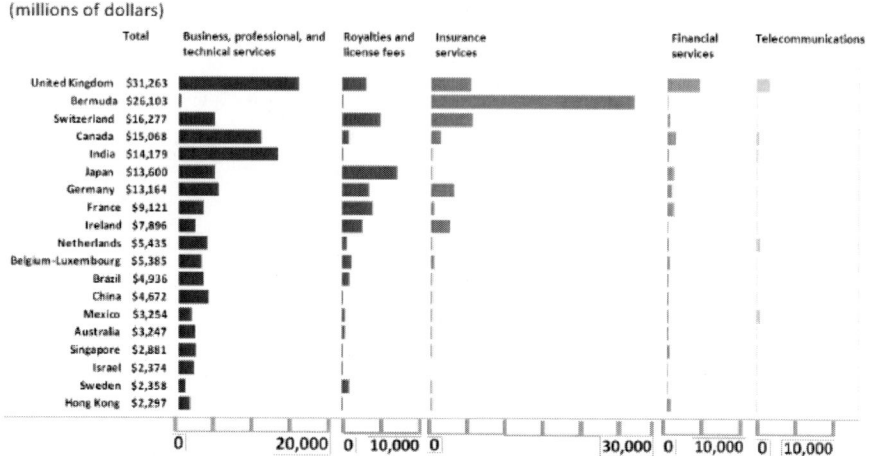

Source: Economics and Statistics Administration analysis using data from the Bureau of Economic Analysis.

Note: Data on imports of insurance services to Israel are suppressed to avoid disclosures of Data if individual companies.

Figure 6. U.S. Imports of Digitally-Deliverable Services by Service Category and Selected.

The top destination countries for U.S. exports of digitally-deliverable services are shown in Figure 5. Trade with the United Kingdom accounted for 10 percent of U.S. digitally-deliverable services exports in 2011. The top six countries to which the U.S. exports these services did not change from 2006 to 2011, although their respective ranks did change. The biggest gainers were China and Brazil. In 2006, these countries ranked 14th and 18th, respectively, in the list of the largest export destinations for U.S. digitally-deliverable services. In 2011, however, these countries ranked 7th and 8th.

By category, the United States primarily exports business, professional, and technical services and royalties and license fees, as previously mentioned.

The United States exports the greatest value of telecommunications services to Brazil (25 percent) and the United Kingdom (15 percent).

Turning to U.S. imports of digitally-deliverable services, from 2006 to 2011 the top ten U.S. trading partners did not change, although the rankings did shift within the top ten (See Figure 6). In 2011, the U.S. imported $9.3 billion more in digitally-deliverable services from India than in 2006, moving the country from 9_{th} place to 5th, overtaking Japan, Germany, France, and Ireland. This $9.3 billion increase accounts for 12.5 percent of the increase in U.S. imports of digitally-deliverable services across all countries during this period. In the Western Hemisphere, outside of South and Central America and Canada, there was a considerable amount of U.S. trade in digitally-deliverable services.[18] In 2011, this region accounted for 8 percent of all exports of this type from the U.S. and 17 percent of all imports. In the insurance services category, this region played a large role in both U.S. exports and imports. In 2011, Bermuda, which identifies itself as "the world's risk capital," was the recipient of 18 percent of U.S. exports and the source of 49 percent of U.S. imports of insurance services.[19,20]

CONCLUSION

Digitally-deliverable services are an important contributor to U.S. trade and help to improve the overall trade balance. When these services are considered as inputs throughout the production process, their value is even more apparent. Virtually no good or service is produced in the United States without input from one of the digitally-deliverable services studied in this report. Although we do not have precise data on the amount of these services that is actually delivered digitally, we do know that digital service delivery is becoming increasingly important.

Even agriculture uses digital services— ranchers are embedding sensors into cattle that transmit data on location and health— and it is estimated that each cow generates 200 megabits of data per year. Healthcare providers diagnose cataracts and skin disorders using pictures sent by patients via mobile phone.[21] Cisco predicts that by 2015, there will be 25 billion connected Internet devices—3.5 times the predicted world population.[22] As people around the world become more connected, industries will continue to find ways to serve customers digitally.

To fully understand how the U.S. economy is adapting to these changes and how to properly implement policies on international trade and other economic matters, the Federal government must continue to adapt and expand its statistics to capture these changes.

ACKNOWLEDGMENTS

The authors would like to thank the following for their contributions to this report:

Economics and Statistics Administration	Sabrina Montes
Bureau of Economic Analysis	Maria Borga
	Gabriel Medeiros
	Matthew Russell
	Sally Thompson

We also appreciate helpful comments we have received from staff in the following agencies and groups:

Economics and Statistics Administration
Bureau of Economic Analysis
Internet Policy Task Force, Department of Commerce
National Telecommunications and Information Administration
United States Patent and Trademark Office
International Trade Administration
Office of Public Affairs, Department of Commerce
United States Trade Representative
Office of Science and Techology Policy, White House

The authors are economists in the Office of the Chief Economist of the U.S. Department of Commerce's Economics and Statistics Administration.

TECHNICAL APPENDIX

Input-Output Analysis

This section describes the methodology used to calculate the value of digitally-deliverable services including inputs throughout the supply chain for the production process for these services. Because BEA's service categories are not based on the industry codes used by Census and for the input-output (I-

O) tables, we developed a concordance (see Table A-1) that matches the BEA categories to the I-O codes used in the annual I-O tables and the 2002 benchmark I-O table. We also included the 2002 North American Industrial Classification System (NAICS) industry codes that correspond to the cross-border trade in services categories and the I-O codes. Because the BEA service categories are broad, they capture several industries that are less likely to be highly digitally-deliverable—for example, 5621 (waste collection) and 8121 (personal care services), as well as many others. However, BEA does not have cross-border trade in services data at a finer level of detail. For consistency across data sources, we included all of these NAICS codes in our analysis.

We used two of the annual I-O tables—the industry-by-commodity total requirements table and the use table—to calculate the percentage of export value contributed by digitally-deliverable services throughout the production process for the years 2002 through 2011.

The step-by-step process below describes how to use the I-O tables to calculate the percentage of U.S. export value contributed by digitally-deliverable services throughout the production process. This process was repeated using annual I-O tables for the years 2002 through 2011.

1. For each row (industry) in the industry-bycommodity table, multiply each column entry (commodity) in the row by the corresponding row entry (commodity) in the exports column of the use table. Sum the results for each row (industry) in the industry-by-commodity table to obtain the gross output of each industry required to produce total U.S. goods and services exports.
2. For each column (industry) in the use table, divide the entry for "Total Value Added" by "Total Industry Output" to obtain the share of value added to output for each industry.
3. For each industry, multiply the result obtained in step 1 by the result obtained in step 2 to obtain the export value added for each industry for the given year.
4. Using the results of step 3, locate the export value added for the digitally-deliverable service industries. The included industries have the following I-O codes: 511, 512, 513, 514, 521CI, 523, 524, 525, 531, 532RL, 5411, 5415, 5412OP, 55, 561, 562, 621, 622HO, 624, 711AS, 713, and 81.
5. Sum the results of step 4 to obtain the total value added by digitally-deliverable services in the export supply chain.

6. Divide the value of digitally-deliverable service inputs obtained in Step 5 by the sum of the value of total inputs obtained in step 3 and multiply by 100. This is the percentage of export value contributed by digitally-deliverable services throughout the production process.

Sensitivity Analysis

As previously mentioned, and as the industry concordance shows, the limited number of BEA cross-border trade-in-services categories leads us to include in our analysis some NAICS industries that are less likely to be digitally-deliverable (having the potential or likelihood to be delivered digitally) or actually delivered digitally. Specifically, the business, professional, and technical services category includes several NAICS industries that are arguably not digitally-deliverable. In an attempt to understand how the portion of exports attributed to digitally-deliverable services, including intermediate inputs, might change if we altered the mix of industries included in our analysis, we conducted a sensitivity analysis using the 2002 benchmark I-O tables. The goal of the sensitivity analysis was to see how the inclusion or exclusion of various industries affected the overall proportion of exports attributable to digitally-deliverable services.

Using the 2002 I-O benchmark tables and the BEA cross-border trade categories, we found that digitally-deliverable services contributed 39.5 percent of the value of all exports that year. Following the methodology described above, we calculated the percentage of export value added by digitally-deliverable services for a subset of the industries included in the BEA categories. We developed two new sets of digitally-enabled services categories; Table A-2 displays the I-O industries included in each of the three groupings: BEA, ESA 1, and ESA 2. For the two new groups, we removed I-O industries that, in our judgment, were unlikely to be highly digitally-deliverable. For example, it is hard to imagine digital delivery of services in the automotive equipment rental and leasing industry. The same is true for offices of physicians, dentists, and other health practitioners. It is important to emphasize that industries were included or not included based solely on our understanding of the services provided and our judgment as to whether or not the service can be delivered, and traded, digitally. Services that may be enhanced or improved by digital processes remain excluded if the ultimate delivery of the service to the customer is not performed digitally.

For ESA 2, we were stricter in our judgment and removed industries that we believe are unlikely to deliver a large portion of their services digitally. For example, although it is likely that industry 621B00 (medical and diagnostic labs and outpatient and other ambulatory care services) provides digital X-ray and other medical imaging services, it is more difficult to imagine outpatient and other ambulatory care services that can be provided digitally. For this reason, we included the industry in the ESA 1 category but excluded it from the ESA 2 category.

Industry 561500 (travel arrangement and reservation services) was not originally included in the BEA digitally-enabled services categories, but we have included it in both the ESA 1 and ESA 2 groups of digitally-deliverable services for the sensitivity analysis. In our judgment, the output of this particular industry, which is included in the travel category of BEA's cross-border trade data, could be and very often is delivered digitally. According to the Census Bureau's E-Stats report, 28 percent of revenues in this industry were from e-commerce sales in 2011.[23]

Table A-3. Results of Sensitivity Analysis, 2002

Industry Group	Share of digitally-deliverable exports (including intermediate inputs)
BEA ESA 1 ESA 2	39.5% 33.8% 20.7%

Source: Economics and Statistics Administration analysis of data from the Bureau of Economic Analysis.

In both of the industry groups we created for the sensitivity analysis, the share of exports accounted for by digitally-deliverable services is lower than in the industry group assembled by BEA (See Table A-3). In large part, this is due to the variety of industries included in business, professional, and technical services. Due to data limitations, BEA's cross-border trade statistics cannot be broken down to a low enough level of detail to match the industries we selected from the I-O tables.

Because the detailed, benchmark I-O tables are only available every five years and have a considerable time lag, it is not possible to perform this analysis on an annual basis. The results of this sensitivity analysis can serve as upper and lower bounds to the actual percentage of exports attributable to digitally-deliverable services. To maintain consistency with the UNCTAD definition for cross-country comparisons, however, the full business, professional, and technical services category may need to be included in any analysis of digitally-deliverable services.

Notes on Other Service Industries

Wholesale trade is not included in the digitally-deliverable services group because of the way trade in this service industry is measured. Cross-border wholesale trade margins are included in the value of goods trade, so they don't appear in the cross-border trade in services data. One exception is if a wholesaler purchases a good from one country and sells it to another country with the good never entering the United States. In this situation, that wholesaler's margin is included in merchanting services and is included in the BEA business, professional, and technical services category (as a trade-related service). As there is no way to parse out which portion of the wholesale trade industry's exports are attributable to this service, it is excluded from our analysis.

Retail trade does not record any exports. Because cross-border trade data are collected at ports, wholesalers and truck services are typically the only parties involved in the transactions. Retailers are not present at the port and there is no retail price associated with exported items. As previously mentioned, any retail transaction costs associated with the exported digitally-deliverable services are included with the value of the exported service in the digitally-deliverable service category.

Table A-1. Industry Concordance of Digitally-Deliverable Services

BEAT,de inSeMeCwgo,y	Inpt-O.tp.tCode (MnI Tble)	Inpt-O.tp.tCode (BenhntTble)	22 NAICSCodes
Business, professional, and tednkaI services (except construction)	511 (part) Publishing industries (indudessoftware)	511110 Newspaperpublisher	5111 Newspaper, periodical, book, and directory publishers
		511120 Periodical publishers	
		511130 Bokpublisher	
		511iA0 Directory, mailing list, and other publishers	
		5161w Internet publishingand broadcasting	5161 Internet publishingand broadcasting
	514 Information and data prcessingservics	5181w Internet service providers and web search portals	51S1 Internet service providers and web search portals
		5182w Data processing, hosting, and related services	51S2 Data processing, hosting, related rvices
		51 Other information services	5191 Other information services

Digital Economy and Cross-Border Trade

BEAT,de inSeMeCwgo,y	Inpt-O.tp.tCode (MnI TbIe)	Inpt-O.tp.tCode (BenhntTbIe)	22 NAICSCodes
	531 Real estate	531 Real estate	5311 Lessorsof real estate
			5312 Offices of real estate agents and brokers
			5313 Activities related to real estate
	532RL(part) Rental and leasingservics and lessors of intangible assets	5321w Automotive equipment rental and leasing	5321 Automotive equipment rental and leasing
		532230 Videotape and disc rental	5322 (part) Consumer goods rental
		532AGeneral and consumergoods rental exctptvideotapes and discs	5322 (part) Consumergoods rental
			5323 General rental centers
		534 Commerdal and industrial madineryand equipment rental and leasing	5324 Commericl and industrial madineryand equipment rental and leasing
	5411 Legalservices	5411w Legal services	5411 Legal services
	54120P Miscellaneous professional, scientific, and technical services	5412w Accounting, tax preparation, bookkeeping, and payroll services	5412 Accounting, tax preparation, bookkeeping, and payroll services
		5413w Arditecthral, engineerin& and related services	5413 Arditecthral, engineering, and related services
		5414w Specialized design services	5414 Specialized design services
		S4lzloManagement, scientific, and tedinical consultingservices	5416 Management, scientific, and technical consultingservicts
		5415A0 Environmental and other technical consulting services	
		5417w Scientific research and development services	5417 Scientific research and development services
		541S Advertising and related services	541 Advertising and related services
		5419A0 All other miscellaneous professional, scientific, and technical services	5419 Other professional, scientific, and tednical services
		541920 Photographicserics	
		541940 Veterinary services	
	S4lsComputersystems design and related services	541511 Custom computerprogrammingservices	5415 Computersystemsdesign and related services

Table A-1. (Continued)

BEAT,de inSeMeCwgo,y	Inpt-O.tp.tCode (MnI TbIe)	Inpt-O.tp.tCode (BenhntTbIe)	22 NAICSCodes
		541512 Computerystemsdesign services	
		54151A Othercomputer related services, induding facilities management	
	55 Management ofcompaniesand enterprises	55 Management of companiesand enterprises	5511 Managementof companies and enterprises
	561 (part) Administrative and supportservices	5611w Office administrative services	5611 Office administrative services
		5612Facilitiessupportservices	561 2 Fadlities supportservices
		5613w Employment services	5613 Employment services
		5614w Business support services	5614 Businesssupportservicts
		5616w Investigation and securityservics	5616 Investigation and securityservics
		5617w Services to buildings and dwellings	5617 Services to buildings and dwellings
		561 Othersupport services	5619 Othersupportservics
	562 Waste managementand remediation services	56W Waste managementand remediation services	5621 Waste collection
			5622 Waste treatment and disposal
			5629 Remediation and otherwaste management servicEs
	621 Ambulatory health care services	621Am Officesofphysidans, dentists, and otherhealth practitioners	6211 Office of physicians
			6212 Offices of dentists
			6213 Offices ofotherhealth practitioners
		621B Medical and diagnostic labsand outpatient and otherambulatorycare services	6214 Outpatientcare centers
			6215 Medical and diagnostic laboratories
			6219 Other ambulatory health care services
		6216w Home health care services	6216 Home health care services
	622110 Hospitals and nursingand residential care facilities	62W HOspitals	6221 General medical and surgical hospitals
			6222 Psydiiatric and substance abuse hospitals
			6223 Specialty (excEpt

BEAT, de inSeMeCwgo,y	Inpt-O.tp.tCode (MnI TbIe)	Inpt-O.tp.tCode (BenhntTbIe)	22 NAICSCodes
			psydiiatric and substance abuse) hospitals
		62Nursingand residentialcare fadlities	6231 Nursingcare facilities
			6232 Residential mental retardation, mental health and substance abuse facilities
			6233 Community care fadlities for the elderly
			6239 Other residential care facilities

*5615 Travel arrangement and reservation services are inlcuded in travel services which is not considered a digitally-deliverable service category

Source: Economics and Statistics Administration and the Bureau of Economic Analysis

BEAT, de inSeMeCwgo,y	Inpt-O.tp.tCode (MnI TbIe)	Inpt-O.tp.tCode (BenhntTbIe)	22 NAICSCodes
Business, pnofessional, and technical senvices (eoceptconstnuction) (continued)	624 Social sistance	624A24 Individual and familysenvices	6241 Individual and familysenvices
		624224 Communityfood, housing, and othen neliefsenvices, including nehabilitation senvices	6242 Communityfood and housing, and emetgencyand othenneliefnvices
			6243 Vocational nehabilitation nvices
		624424 Child daycane senvices	6244 Child daycane senvices
	71241 Petfonningants, specoatonspotto, moeums, and nelated activities	711124 Petfonningants companies	7111 Pfonming companies
		711224 Specoatonspotto	7112 Spectatonspotts
		712424 Pnomotes of penfonning atts and sponts and agents fon public figunes	7113 Pnomotets of penfonmingatts, sponts, and si,nilanevents
			7114 Agents and managensfonattists, athletes, entettainens, and othen public figunes
		711524 Independentantists, wtitets, and penfoninets	7115 Independentattists, wnitens, and petfonnets
		712Museums, histonical sites, zoos, and paths	7121 Museums, histotical sites, and simnilaninstitutions
	713 Amusements, gemnbling, and neoeation industnies	713A24 Amusement paths, ancades, and gamnbling industties	7131 Amnusemnent paths and ancades
			7132 Gambling industties
		713640 Otheramusement and necneation industties	7139 Othenamusementand neoceation industties
		713940 Fitness and necneational spottscentets	

Table A-1. (Continued)

BEAT,de inSeMeCwgo,y	Inpt-O.tp.tCode (MnI TbIe)	Inpt-O.tp.tCode (BenhntTbIe)	22 NAICSCodes
		713950 Bowlingcentets	
	81 Other sencices, escept govemment	8111A0 Automotive nepainand maintenance,esceptcanwashes	8111 Automotive nepainand maintenance
		811192 Canwashes	
		811224 Becononicand pnecision equipment nepainand maintenance	8112 Electnonicand pmcision equipment mpainand maintenance
		811324 Commenoal and mndustnal machmnenyand equipment nepamnand maintenance	8113 Commendal and industtial mad,inenf and equipment (exceptautomotive and elecononic) mpainand maintenance
		811424 Petsonal and household goods mpainand maintenance	8114 Pensonal and houshold good mpainand maintenance
		812124 Petsonal cam senvices	8121 Pensonal cane senvices
		812224 Death cane senvices	8122 Death cam senvices
		812324 Dnf-deaning and laundnysenvices	8123 Dnfdeaning and laundnysenvices
		812Othenpetsonalsenvices	8129 Othenpetsonalsencices
		813124 Religious ongenizations	8131 Religiousotganiaations
		813A24 Gtantmaking, giving, and sodal advocacy otganiaations	8132 Gtantmaking and giving senvices
			8133 Social advocacy ongenizations
		813640 Civic, social, pnofessional, and similanotganiaations	8134 Civic and sodal otganiaations
			8139 Business, pnofessional, labon, political, and similanotganiaations
		814Ptivate households	8141 Pnivate households
Royaltiesand license fees	511 (pad) Publishing industties (indudessoftwam)	SSS224Softwam published	5112 Softwam published
	512 Motion picoum and sound	512140 Motion picture and video industties	5121 Motion pictum and video industnies

Digital Economy and Cross-Border Trade 79

BEAT,de inSeMeCwgo,y	Inpt-O.tp.tCode (MnI TbIe)	Inpt-O.tp.tCode (BenhntTbIe)	22 NAICSCodes
	mcondingindustnies	512240 Sound mconding industties	5122 Sound seconding industnies
	532RL(patt) Rental and leasingsenvices and lessotsof intangible assets	533Lessots of nonfinancial intangible assets	5331 Lessonsof nonfinandal intangible assets (eoceptcopynighted wombS)
Insutance senvices	524 Insurance camiets and mlated activities	524124 Instance cantiens	5241 Insurance camiets
		524224 Instance agencies, bnoketages, and mlated activities	5242 Agencies, bnokenages, and othen insutance mlated activities
Financialsenvices	52501 Fedetal Reserve banks,cmditintennediation,and mlatedactivities	52AMonetanyauthonitiesa nddepositonfcmditintennediation	5211 Monetanyauthonities-centnal bank
			5221 Depositonycnedit intennediation
		522495 Nondepositonycmdit intennediation and mlated activities	5222 Nondepositonycmdit intennediation
			5223 Activities mlated to cmdit intennediation
	523 Secutities,commodityco ntnacos,and investments	523Seomnities,commodity conttacts,investments,and mlatedacoivities	5231 Secutitiesandcommodityc ontnactsintennediation and bnokenage
			5232 Secutities and commodity esdianges
			5239 Othenfinancial investment activities
	525 Funds, tnmsts, and othenfinandal vehides	525Funds, tnmsts, and othenfinancial vehicles	5251 Insunance and employee benefitfunds
			5259 Otheninvestment pools and funds
Telecommunications	513 Bnoadcastingand telecommunications	515124 Radioand television bnoadcasting	5151 Radioand television bnoadcasting
		515224 Cable and othensubsctiption pnognamming	5152 Cable and othensubsctiption pnogtamming
		5S7 Telecommunications	5171 Wised telecommunications cafflets
			5172 Wimlesstelecemmunicationscamiets (eoceptsatellite)
			5173 Telecommunications nesellets
			5174 Satellite telecommunications
			5175 Cable and other pnognam distnibution

Table A-1. (Continued)

BEAT,de inSeMeCwgo,y	Inpt-O.tp.tCode (MnI TbIe)	Inpt-O.tp.tCode (BenhntTbIe)	22 NAICSCodes
			5179 Other telecommunications

Source: Economics and Statistics Administration and the Bureau of Economic Analysis.

Table A-2. Digitally-Deliverable Service Industries for Sensitivity Analysis

BEA Trade in Services Category	I-O Industry Code and Title		BEA	ESA 1	ESA 2
Business, professional, and technical services (except construction)	511110	Newspaper publishers	✓	✓	✓
	511120	Periodical publishers	✓	✓	✓
	511130	Book publishers	✓	✓	✓
	5111A0	Directory, mailing list, and other publishers	✓	✓	✓
	516110	Internet publishing and broadcasting	✓	✓	✓
	518100	Internet service providers and web search portals	✓	✓	✓
	518200	Data processing, hosting, and related services	✓	✓	✓
	519100	Other information services	✓		
	531000	Real estate	✓		
	532100	Automotive equipment rental and leasing	✓		
	532230	Video tape and disc rental	✓		
	532A00	General and consumer goods rental except video tapes and discs	✓		
	532400	Commercial and industrial machinery and equipment rental and leasing	✓	✓	
	541100	Legal services	✓	✓	
	541200	Accounting, tax preparation, bookkeeping, and payroll services	✓	✓	
	541300	Architectural, engineering, and related services	✓	✓	✓
	541400	Specialized design services	✓	✓	✓
	541511	Custom computer programming services	✓	✓	✓
	541512	Computer systems design services	✓	✓	✓
	54151A	Other computer related services, including facilities management	✓	✓	✓
	541610	Management, scientific, and technical consulting services	✓	✓	✓
	5416A0	Environmental and other technical consulting services	✓	✓	✓

BEA Trade in Services Category	I-O Industry Code and Title		BEA	ESA 1	ESA 2
	541700	Scientific research and development services	✓	✓	✓
	541800	Advertising and related services	✓	✓	✓
	5419A0	All other miscellaneous professional, scientific, and technical services	✓	✓	✓
	541920	Photographic services	✓		
	541940	Veterinary services	✓	✓	
	550000	Management of companies and enterprises	✓	✓	
	561100	Office administrative services	✓		
	561200	Facilities support services	✓	✓	
	561300	Employment services	✓	✓	
	561400	Business support services	✓		
	561600	Investigation and security services	✓		
	561700	Services to buildings and dwellings	✓		
	561900	Other support services	✓		
	562000	Waste management and remediation services	✓		
	621A00	Offices of physicians, dentists, and other health practitioners	✓		
	621B00	Medical and diagnostic labs and outpatient and other ambulatory care services	✓		
	621600	Home health care services	✓		
	622000	Hospitals	✓		
	623000	Nursing and residential care facilities	✓		
	624A00	Individual and family services	✓		
	624200	Community food, housing, and other relief services, including rehabilitation services	✓		
	624400	Child day care services	✓		
	711100	Performing arts companies	✓		
	711200	Spectator sports	✓		
	711A00	Promoters of performing arts and sports and agents for public figures	✓		
	711500	Independent artists, writers, and performers	✓		
	712000	Museums, historical sites, zoos, and parks	✓		
	713A00	Amusement parks, arcades, and gambling industries	✓	✓	
	713B00	Other amusement and recreation industries	✓	✓	
	713940	Fitness and recreational sports centers	✓	✓	
	713950	Bowling centers	✓	✓	

Table A-2. (Continued)

BEA Trade in Services Category	I-O Industry Code and Title		BEA	ESA 1	ESA 2
	8111A0	Automotive repair and maintenance, except car washes	✓		
	811192	Car washes	✓		
	811200	Electronic and precision equipment repair and maintenance	✓		
	811300	Commercial and industrial machinery and equipment repair and maintenance	✓		
	811400	Personal and household goods repair and maintenance	✓		
	812100	Personal care services	✓		
	812200	Death care services	✓		
	812300	Dry-cleaning and laundry services	✓		
	812900	Other personal services	✓		
	813100	Religious organizations	✓		
	813A00	Grantma king, giving and social advocacy organizations	✓		
	813B00	Civic, social, professional, and similar organizations	✓		
	814000	Private households	✓		
Royalties and license fees	511200	Software publishers	✓	✓	✓
	512100	Motion picture and video industries	✓	✓	✓
Insurance services	524100	Insurance carriers	✓	✓	✓
	524200	Insurance agencies, brokerages, and related activities	✓	✓	✓
Financial services	52A000	Monetary authorities and depository credit intermediation	✓	✓	✓
	522A00	Nondepository credit intermediation and related activities	✓	✓	✓
	523000	Securities, commodity contracts, investments, and related activities	✓	✓	✓
	525000	Funds, trusts, and other financial vehicles	✓	✓	✓
Telecommu-nications	515100	Radio and television broadcasting	✓	✓	✓
	515200	Cable and other subscription programming	✓	✓	✓
	517000	Telecommunications	✓	✓	✓
Travel*		Travel arrangement and reservation services		✓	✓

*Other I-O Codes are included in the the travel category, but are not listed here since they are not included in our analysis.

End Notes

[1] For consistency with the input-output tables, Table 1 in this report excludes re-exports and presents a larger number (20 percent) for the digitally-deliverable share of total exports. In contrast, official statistics include re-exports.

[2] BEA will introduce new trade in services categories based on new international standards with their June 2014 revision. The new standards require differentiation of licenses to use from licenses to reproduce and distribute. The two manuals that give guidance on trade in services are published by the International Monetary Fund (http://www.imf.org/external/pubs/ft/bop/2007/bopman 6.htm) and the United Nations Statistical Division (http://unstats.un.org/unsd/tradeserv/tfsits/msits2010.ht m).

[3] Services supplied through affiliates cover the delivery of services to international markets through establishment of a commercial presence. While an important mode of delivery, a discussion of services supplied through affiliates is beyond the scope of this report. Data on services supplied through affiliates are available from BEA at: http://www.bea.gov/international/internationalservices.htm #detailedstatisticsfor1.

[4] http://www.bea.gov/international/pdf/trendsin digitallyenabledservices.pdf.

[5] This paper's approach excludes services that are not primarily delivered online. For example, education is a services trade category for which the primary mode of delivery is in-person rather than digital; therefore, the education services category is not included in digital trade. Distance learning, which is primarily delivered online, is not part of the education services trade category and is captured in official statistics as training, a component of business, professional, and technical services.

[6] For consistency with the input-output tables, Table 1 in this report excludes re-exports and presents a larger number (20 percent) for the digitally-deliverable share of total exports. Official statistics include re-exports and should be used wherever possible.

[7] Cross-border trade data for 2012 were available at the time this report was written. However, to be consistent with the input-output data available, we opted to use 2011 data.

[8] See http://stats.oecd.org/Index.aspx?DataSetCode=TIVAOEC DWTO. The indicator SERV_VAGR—"Services value added embodied in gross exports by source country, as % of gross exports"—displays the relevant statistic.

[9] We also performed a more fine-grained analysis using the 2002 benchmark I-O tables, which present industries at a much more detailed level than the annual tables. Our result—39.5 percent—was similar to the result of the annual analysis. BEA will release new benchmark I-O tables for 2007 and new annual tables for 2012 in December 2013, at which point it will be possible to update this analysis.

[10] The U.S. Census Bureau collects data on revenues in services industries in the Service Annual Survey. Data on e-commerce revenues are only available for limited years due to various factors, including changes in industry classification and changes to the scope of the survey. "E-commerce" describes on-line transactions whether over open networks such as the Internet or proprietary networks running systems such as Electronic Data Interchange (EDI). http://www.census.gov/econ/estats/2011/all2011tables.html and http://www.census.gov/econ/estats/about.html.

[11] For a detailed discussion of the increasing use of digital technologies in these services, see "Digital Trade in the U.S. and Global Economies, Part 1." United States International Trade Commission Publication 4415, July 2013. Pages 3-7 through 3-24. Accessed online September 9, 2013. http://usitc.gov/publications/332/pub4415.pdf.

[12] "IFPI Digital Music Report 2012 Engine of a Digital World" Accessed online December 18, 2013. http://www.ifpi.org/content/library/DMR2013.pdf.
[13] "Counting the Change." The Economist, August 17, 2013, page 53.
[14] "Omnipotent, or omnishambles?" The Economist, August 3, 2012, page 53.
[15] Within the "Europe" category are separate estimates for Belgium-Luxembourg, France, Germany, Ireland, Italy, Netherlands, Norway, Spain, Sweden, Switzerland, United Kingdom, and an "Other" category covering the remainder of Europe.
[16] Within the "Asia and Pacific" category are separate estimates for Australia, China, Hong Kong, India, Indonesia, Japan, Republic of Korea, Malaysia, New Zealand, Philippines, Singapore, Taiwan, Thailand, and an "Other" category covering the remainder of the region.
[17] Within the "South and Central America" category are separate estimates for Argentina, Brazil, Chile, Mexico, Venezuela, and an "Other" category covering the remainder of the region.
[18] Within the "Other Western Hemisphere" category, Bermuda is the only country specifically broken out due to the higher volumes of trade that occur between Bermuda and the United States compared to other countries in the region.
[19] http://www.bermuda-insurance.org/
[20] See also: United States International Trade Commission. Property and Casualty Insurance Services: Competitive Conditions in Foreign Markets, Box 3.4 Bermuda's International Insurance Industry. USITC Publication 4068. March 2009. Accessed online September 4, 2013. http://www.usitc.gov/publications/332/pub4068.pdf.
[21] "The Next Wave of Digitization Setting Your Direction, Building Your Capabilities." Booz & Company. Accessed online August 22, 2013. http://www.booz.com/media/uploads/BoozCo-NextWave-of-Digitization.pdf.
[22] "The Internet of Things: How the Next Evolution of the Internet is Changing Everything." Cisco White Paper. Accessed online December 12, 2013. http://www.cisco.com/web/about/ac79/docs/innov/IoTI BSG0411FINAL.pdf.
[23] http://www.census.gov/econ/estats/2011/all2011tables.html, Table 3.

In: U.S. Technological Endeavors
Editor: Beverly Howard

ISBN: 978-1-53610-547-6
© 2017 Nova Science Publishers, Inc.

Chapter 3

U.S. SEMICONDUCTOR MANUFACTURING: INDUSTRY TRENDS, GLOBAL COMPETITION, FEDERAL POLICY*

Michaela D. Platzer and John F. Sargent Jr.

SUMMARY

Invented and pioneered in the United States shortly after World War II, semiconductors are the enabling technology of the information age. Because of semiconductors new industries have emerged and existing ones, such as aerospace and automotive, have been transformed. Semiconductors have contributed in powerful and unique ways to nearly all fields of science and engineering, and semiconductors' economic and military importance has made the industry's health a focus of congressional interest for nearly 70 years. In July 2015, Congress formed the Semiconductor Caucus, a group that seeks to advance policies that support the U.S. semiconductor industry.

The federal government played a central role in the creation of the U.S. semiconductor industry. World War II funding for electronics and materials research and development (R&D) provided essential support for the invention and refinement of semiconductors. Federal investments in computing advances also created an important application for semiconductors and federal acquisitions for defense, space, and civilian

---

* This is an edited, reformatted and augmented version of a Congressional Research Service publication, R44544, dated June 27, 2016.

applications made up the lion's share of the early semiconductor market. In the face of formidable competition from Japanese companies in the 1980s, Congress co-funded SEMATECH, an industry research consortium devoted to developing the technologies needed by U.S. firms to remain competitive. Today, Congress continues to provide funding for R&D and development of scientific and engineering talent in support of the industry. In 2015, Congress acted to make the R&D tax credit permanent, a policy priority of the industry.

An ongoing issue of congressional interest is the retention of high-value semiconductor manufacturing in the United States. In 2015, semiconductor manufacturers directly employed 181,000 workers, who earned an average wage of $138,100, more than twice the average wage for all U.S. manufacturing workers. Increasingly, however, U.S. firms are building semiconductor fabrication plans (fabs) abroad, primarily in Asia. In addition, some semiconductor firms are going "fab-less," focusing corporate resources on chip design and relying on contract fabs abroad to manufacture their products. At year-end 2015, there were 94 advanced fabs in operation worldwide, of which 17 were in the United States, 71 in Asia (including 9 in China), and 6 in Europe. The Chinese government regards the development of a domestic, globally competitive semiconductor industry as a strategic priority with a stated goal of becoming self-sufficient in all areas of the semiconductor supply chain by 2030. China faces significant barriers to entry in this mature, capital-intensive, R&D-intensive industry.

Because the primary market for U.S.-based semiconductor firms is located outside the United States (83% in 2015), passage of the Trans-Pacific Partnership (TPP) agreement and successful conclusion of the ongoing Transatlantic Trade and Investment Partnership (TTIP) negotiations with Europe are top industry priorities. In 2015, exports of U.S. semiconductors and related devices totaled $41.8 billion, making it the nation's fourth-largest overall exporting industry. The 2015 expansion of the World Trade Organization (WTO) Information Technology Agreement (ITA), a plurilateral tariff-cutting agreement focused on trade in information technology goods, is considered a major success for the U.S. semiconductor industry.

Semiconductor manufacturing also raises national security concerns, including secure access to trusted suppliers of advanced semiconductors and other critical technology components that are important for certain defense and national security applications. The House Armed Services Subcommittee on Oversight and Investigations held a hearing on this issue in October 2015.

INTRODUCTION

Semiconductors, tiny electronic devices based on silicon or germanium, provide data processing capabilities in millions of products, from coffee pots to space vehicles. The U.S. government played a significant role in the development of semiconductor technology, and domestic research and production have long been matters of intense congressional interest.

U.S.-headquartered semiconductor firms accounted for about half of worldwide semiconductor sales in 2015.[1] However, U.S.-headquartered producers face stiff competition from firms headquartered in South Korea, Japan, and Taiwan; moreover, the Chinese government has identified global leadership in semiconductors as a national priority. Further, the United States accounts for a diminishing share of global semiconductor production capacity, as manufacturers establish plants in locations where generous subsidies are available or customers in user industries, such as electronic products manufacturing, are nearby. In July 2015, Members of Congress concerned about the industry's competitiveness formed a Semiconductor Caucus to support increased federal funding for semiconductor research activities, among other objectives.

SEMICONDUCTOR INDUSTRY BASICS

A semiconductor chip (also known simply as a "semiconductor" or "chip") is a tiny electronic device (generally smaller than a postage stamp) comprised of billions of components that store, move, and process data. These functions are made possible by the unique properties of semiconducting materials, such as silicon and germanium, which allow for the precise control of the flow of electrical current.

Semiconductors are the enabling technology of the information age. Semiconductors allow computers to run software applications, such as email, Internet browsers, and word processing and spreadsheet programs and to store documents, photographs, videos, music, and other data. They also provide the "brains," memory, and data communication capabilities of countless other products, from cell phones and gaming systems to aircraft and industrial machinery to military equipment and weapons. Even many products with roots in mechanical systems are now heavily dependent on chip-based electronics: one car manufacturer asserts that some of its models incorporate as many as

6,000 semiconductors.[2] And one expert on software in cars estimates that premium-class automobiles can contain close to 100 million lines of software code (instructions) that the chips use to control the vehicle.[3]

Semiconductor History and Technological Challenges

Military applications were the primary driver for the invention of semiconductors. Early computers relied on thousands of vacuum tubes, crystal diodes, relays, resistors, and capacitors to perform simple calculations. The federal government, academia, and U.S. industry undertook efforts to reduce and simplify the number of these devices. The invention of the transistor, a simple semiconductor device capable of regulating the flow of electricity, was followed by the development of the integrated circuit (IC), in 1958. ICs allowed thousands of resistors, capacitors, inductors, and transistors to be "printed" and connected on a single piece of semiconductor material, so that they functioned as a single integrated device. In addition to funding academic and industrial research that contributed to the early development of semiconductor technology, the federal government played a central role in the commercialization of the technology through purchases of semiconductors for a variety of military, space, and civilian applications.

The semiconductor industry has a rapid internal product development cycle, first described by the former CEO and co-founder of Intel Corporation, Gordon Moore. Moore's Law, which is actually an observation about the pace of development and cost reduction in chip speeds, has held true for decades. It states that the number of transistors in a dense integrated circuit will double about every 18 months to two years, making semiconductors smaller, faster, and cheaper.[4] The effects of Moore's law are evident in short product life-cycles, requiring semiconductor manufacturers to maintain high levels of research and investment spending. A main challenge for the industry is that semiconductor inventory and technology can become obsolete quickly, leaving producers with serious financial problems if they have unsalable inventories as improved designs displace existing products.

A major question facing semiconductor manufacturers is whether fundamental physical limits may soon make it difficult to pack more transistors onto a silicon device in an economical way.[5] If this proves to be the case—the continuing validity of Moore's law is hotly debated—then manufacturers would need to find other methods of improving semiconductors.[6] Research is underway into new approaches to computing (such as quantum computing, optical computing, and neuromorphic (brain-like) computing) that could, theoretically, vastly surpass the storage, processing, and transmission capabilities of semiconductor technology.[7] These approaches, however, face substantial technological obstacles to their realization.

THE GLOBAL SEMICONDUCTOR INDUSTRY

The semiconductor industry is generally characterized by large fluctuations in product supply and demand, depending heavily on the strength of the global economy.[8] U.S.-headquartered firms have the largest share of the global market, measured by sales, at close to 50%.[9] Half of the 20 largest semiconductor firms by revenue in 2015 are headquartered in the United States: Intel, Qualcomm, Micron, Texas Instruments, Broadcom, Apple, SanDisk, NVIDIA, Advanced Micro Devices, and On Semiconductor.[10] Other leading firms are based in South Korea, Japan, Taiwan, and Europe. There are no China-based semiconductor firms on the top 20 list.

Only a handful of companies have the sales volume to operate as integrated device manufacturers (IDMs) operating their own fabrication facilities (known as fabs).[11] Other chip firms are "fabless," meaning that they design and market semiconductors but contract production to "foundries" that manufacture semiconductors to order. Taiwan Semiconductor Manufacturing Company (TSMC), a Taiwanese-headquartered company, operates the world's largest foundry. Fabless semiconductor firms generally enjoy higher and less volatile profit margins than semiconductor manufacturers with integrated operations.[12] Potential risks associated with the use of a contract foundry include availability of capacity, timeliness of production, and quality control.

Semiconductor Industry Sales

Worldwide semiconductor sales reached $335 billion in 2015, up 15.0% over 2012, according to figures from World Semiconductor Trade Statistics (WSTS).[13] During the same period, sales of U.S.-based semiconductor manufacturers rose 14.6%. According to Semiconductor Industry Association (SIA) data, global semiconductor sales have increased at a compounded annual rate of 9.5% over the past 20 years.[14]

In recent years, semiconductor sales of U.S.-based companies have accounted for about half of worldwide semiconductor sales (see *Figure 1*). In 2015, total sales of U.S.-headquartered semiconductor firms experienced a contraction, and its global market share dropped two percentage points to 49.6%. In 2015, U.S.-headquartered firms posted sales of $166 billion.[15] WSTS forecasts a modest increase in worldwide semiconductor industry sales to $347 billion (+4%) in 2017.[16] According to semiconductor industry experts, it seems likely that the U.S. market share will remain around 50% in 2017.

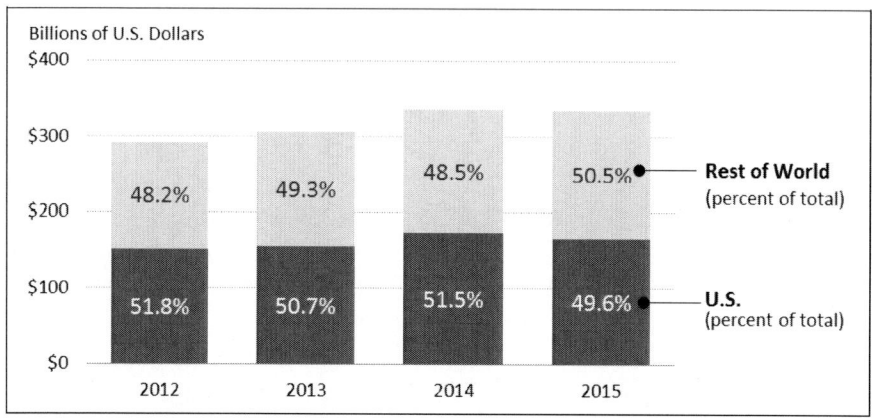

Source: Semiconductor Industry Association (SIA), The U.S. Semiconductor Industry, 2014, 2015, and 2016 Factbooks.

Figure 1. Worldwide and U.S. Semiconductor Industry Sales.

Major Industry Segments

Semiconductors are classified into major product groups, mainly based on their function. Some of these products have broad functionality; others are designed for specific uses. According to SIA, integrated circuits, which are directly embedded onto the surface of the semiconductor chip, account for the overwhelming majority of industry sales (82% in 2015). The remaining 18% of the market is made up of sales in the optoelectronics, sensors, and discretes (O-S-D) market. Optoelectronics and sensors are mainly used for generating or sensing light, for example, in traffic lights or cameras, and discretes are used in electronic devices to control electric current.[17]

Within the integrated circuit market, the four largest segments in 2015 were:

1. *Logic Devices.* Logic devices are used for the interchange and manipulation of data in computers, communication devices, and consumer electronics.[18] Logic devices are the largest category by sales, accounting for 27% of the total semiconductor market.
2. *Memory Devices.* Memory devices store information. This segment includes dynamic random access memory (DRAM), a common and inexpensive type of memory used for the temporary storage of information in computers, and flash memory, which retains data in the absence of a power supply. Memory devices account for 23% of semiconductor market sales.
3. *Microprocessors.* Microprocessors execute software instructions to perform a wide variety of tasks such as running a word processing program or video game. They make up about 18% of semiconductor sales.
4. *Analog Devices.* Analog devices include analog signal processing technologies, data converters, amplifiers, and radio frequency integrated circuits. These devices, for example, convert analog signals like a musical recording on a phonograph into digital signals like a musical recording on a compact disc. Analog device products account for about 13% of semiconductor industry sales.[19]

Many manufacturers specialize in certain types of semiconductors. For example, South Korean manufacturers Samsung and SK Hynix and U.S.-based Micron together account for 90% of global DRAM sales.[20] Heavy dependence on the DRAM market has been a challenge for these companies, as weak demand or excess capacity have at times led to dramatic reductions in prices.[21] U.S.-based Intel Corporation, the largest semiconductor manufacturer by sales, is highly dependent on supplying microprocessors to the personal computer industry. Microprocessors are harder to manufacture, more technologically advanced, and more expensive than other semiconductor products, providing Intel some shelter from competition, but the company is nonetheless affected by weakening global demand for personal computers.[22]

Multicomponent semiconductors (MCOs) represent a fast-growing segment of the semiconductor industry. These devices combine two semiconductors into a single unit, which takes up less room within the finished product and use less power. MCOs are commonly used in smartphones, tablets, and automotive braking, steering, and air bag systems. Although SIA does not track sales figures for this market, the U.S. International Trade Commission (USITC) estimates that MCOs account for between 1.5% and

3.0% of global semiconductor industry sales.[23] Demand growth is expected to be high in coming years as end-use producers use MCOs to make smaller, lighter, and faster devices that consume less power. U.S.-headquartered companies such as Intel, Texas Instruments, Qualcomm, and Broadcom are among the leaders in this market segment.

A few semiconductor companies manufacture mainly for a single buyer. For example, Kokomo Semiconductors, now part of General Motors (GM) Components Holdings, operates a small fab plant in Indiana, where it produces custom integrated circuits for GM.[24] According to industry experts, small semiconductor firms can compete effectively with larger ones by producing specialized chips for particular market niches or by developing new applications for their customers.[25]

Semiconductor Manufacturing

The production of semiconductors is extremely complex, requiring high levels of automation. As semiconductors become smaller and are more densely packed with transistors, the complexity of manufacturing increases.

Figure 2 depicts a simplified schematic of the semiconductor production process. The process has three distinct components:

1. design;
2. front-end fabrication, in which "fabs" create microscopic electric circuits on silicon wafers;[26] and,
3. back-end testing, assembly, and packaging, in which wafers are sliced into individual semiconductors, encased in plastic, and put through a quality-control process.

The majority of design work, performed by computer engineers, now occurs in the United States.[27] The designs are then placed on a wafer of silicon or other material in a sequence of more than 250 photographic and chemical processing steps using equipment produced by firms such as Applied Materials, ASML Holdings, and Lam Research.[28] This front-end fabrication process typically takes about 2 months.[29] Around 87% of advanced worldwide fab capacity is now located outside the United States (see *Table 1*). Back-end production is where chips are assembled into finished semiconductor components and tested for defects. This stage of the manufacturing process is the most labor-intensive and is often performed in countries such as China and

Malaysia, where labor costs are lower than in the United States, Japan, and Europe. The final stage of manufacturing involves the installation of the chips into consumer goods.

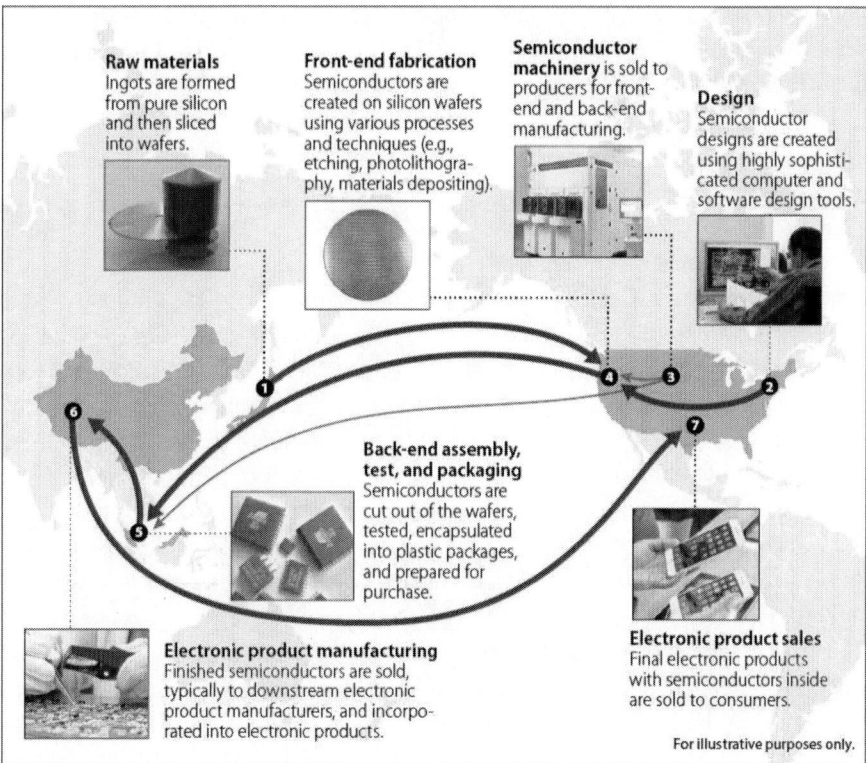

Source: CRS, adapted from information provided by SIA.
Notes: This diagram is for illustrative purposes only. Numbered circles do not necessarily reflect where specific production, services, or sales take place.

Figure 2. Typical Global Semiconductor Production Pattern.

THE U.S. SEMICONDUCTOR MANUFACTURING INDUSTRY

Nationally, there were about 820 firms involved in semiconductor and related device manufacturing in 2013.[30] The U.S. semiconductor industry's contribution to the U.S. economy measured by value added was $27.2 billion in 2014, accounting for approximately one percent of U.S. manufacturing value added.[31] Declining chip prices remain a challenge for semiconductor manufacturers as producers can continually manufacture more powerful chips that contain more functionality at lower prices and the price of semiconductors has fallen consistently over time. For example, according to the Bureau of Labor Statistics (BLS) producer price index, a measure of price changes by industry, semiconductor prices, adjusted for quality and performance, decreased by 46% between 2005 and 2015.[32] Consequently, to maintain or grow their revenue, chip producers must find new markets for their products.

R&D Spending

Because of the constant pressure to innovate, semiconductor manufacturers invest heavily in R&D. According to SIA, industry-wide investment rates in R&D range between 15-20% of sales.[33] In 2012, U.S. semiconductor manufacturers devoted 19.4% of their domestic sales to R&D, which was higher than other large industrial sectors, including pharmaceuticals and medicines and computers and electronic products, based on the most recent available statistics from the National Science Foundation (NSF).[34] By comparison, R&D intensity for all manufacturing industries was 3.6% in 2012. According to an analysis of R&D expenditures by the SIA, R&D performed by semiconductor firms tends to be consistently high, regardless of cycles in annual sales.[35]

In December 2015, a long-standing tax issue for the industry was resolved when Congress made the research and experimentation tax credit (widely referred to as the R&D tax credit) permanent, rather than expiring periodically, as it had over the last few decades.[36] Semiconductor industry lobbyists asserted that making the R&D tax credit permanent will encourage semiconductor companies (and other manufacturers) to plan sustained, long-term R&D efforts.[37]

Employment

According to BLS, the U.S. semiconductor and related device manufacturing industry employed 180,700 workers in 2015, down 38% from 292,100 in 2001.[38] This represented 1.5% of total manufacturing employment in the United States in 2015. The semiconductor manufacturing workforce earned an average wage of $138,100 in 2015, more than twice the average for all U.S. manufacturing workers ($64,305).[39] These employment and wage figures do not include all workers in the industry, as BLS counts employees of fabless semiconductor firms as wholesale trade workers rather than manufacturing workers.[40]

In 2015, nearly one-quarter of all domestic semiconductor manufacturing jobs were in California. Other states accounting for substantial shares of total U.S. semiconductor manufacturing employment include Texas, Oregon, Arizona, Massachusetts, Idaho, and New York.[41]

Semiconductor Manufacturing Locations

Semiconductor manufacturing is highly capital intensive. To produce each new generation of chips, and to benefit from the cost advantage offered by larger wafers, manufacturers must invest in new facilities and equipment and reinvest in existing facilities and equipment. A state-of-the-art plant to make 300-millimeter (12-inch) wafers, the size that allows maximum production efficiency, can cost as much as $10 billion.[42] Between 2011 and 2014, the U.S. Census Bureau reports, the sector's domestic expenditures for new plants and equipment ranged from a low of $17 billion in 2013 to a high of around $22 billion in 2011.[43]

In 2015, about three-fourths of the world's 300mm semiconductor fabrication capacity was located in South Korea, Taiwan, Japan, and China. By comparison, North America (mainly the United States) accounted for about 13% of worldwide 300mm wafer fabrication production capacity in 2015 (see *Table 1*).[44]

Table 1. Global Wafer Fabrication Capacity
300mm Equivalent Wafer Capacity by Country/Region

Country/Region	2015
South Korea	26%
Taiwan	24%
Japan	18%
North America	13%
China	8%
Europe	3%
Rest of World (ROW)	9%

Source: IC Insights, Global Wafer Capacity 2016-2020, http://www.icinsights.com/services/global-wafer-capacity/ report-contents/.

In a 2012 report, the National Academy of Sciences noted the share of total worldwide fabrication capacity located in the United States had dropped from 42% in 1980, to 30% in 1990, and to 16% in 2007. The reasons behind the shift, according to industry experts, were the rapid expansion of Asian semiconductor companies and offshore investment by U.S. companies.[45]

U.S.-headquartered semiconductor fabrication companies conduct more than half of their front-end wafer processing operations in the United States. *Table 2* lists U.S. semiconductor fabs capable of producing 300mm silicon wafers. According to one recent study, semiconductor producers base site selection decisions on tax advantages, supply of engineering and technical talent, quality of water supply, reliability of utilities, environmental permitting process and other regulations, cost of living for employees, and legal protection of intellectual property.[46]

Table 2. 300mm (12-inch) Semiconductor Fabs in the United States, 2015

Company	Number of Facilities	Location	Products
Intel Corporation	2	Chandler, AZ	Logic/Microprocessor Unit (MPU)
Micron Technology	1	Boise, ID	Memory/DRAM/Flash-3D NAND
Intel Corporation	1	Albuquerque, NM	Logic/MPU
GlobalFoundries	2	Malta, NY	Foundry/Dedicated
GlobalFoundries	1	East Fishkill, NY	Foundry/Dedicated
Intel Corporation	4	Hillsboro, OR	Logic/MPU
Samsung	1	Austin, TX	Foundry/System Large Scale Integration (LSI)
Texas Instruments	1	Richardson, TX	Analog/Linear
Texas Instruments	1	Dallas, TX	Analog/Mixed Signal
Micron Technology	1	Lehi, UT	Memory/Flash
Micron Technology	2	Manassas, VA	Memory/DRAM

Source: Semiconductor Equipment and Materials International (SEMI), an industry trade group that represents the manufacturers of semiconductor and flat panel display equipment and materials, provided CRS with a list of semiconductor fabrication plants by email on March 18, 2016 from its Fab Construction database.

Notes: Silicon wafers are available in a variety of sizes. The size of the wafer is an important element in semiconductor manufacturing because the number of chips per wafer increases dramatically as the wafer size increases. State-of-the-art fabs produce silicon wafers that are 300mm in width. Moving from a 200mm wafer to a 300mm wafer increases the number of semiconductor chips by a factor of 2.25 times.

Intel conducts 70% of its wafer fabrication in the United States, at facilities in Arizona, New Mexico, and Oregon.[47] Micron is the only DRAM manufacturer that has factories in the United States, with facilities in Idaho, Utah, and Virginia.[48] Texas Instruments has manufacturing facilities in Maine and Texas.[49] Global Foundries, a company based in California but controlled by the Emirate of Abu Dhabi, has acquired U.S. fabs formerly owned by Advanced Micro Devices and IBM Corporation. All of these companies also manufacture overseas.

Most new semiconductor manufacturing capacity is located outside the United States. According to the Semiconductor Equipment and Materials International (SEMI), an industry trade group that represents the manufacturers of semiconductor and flat panel display equipment and materials, of the 36 new fab projects of all sizes planned to be built worldwide between 2015 and 2017, five were planned for the United States, compared to 14 in China. The other projects will be in Southeast Asia (6), Taiwan (6), Japan (2), Europe (2), and South Korea (1).[50]

International Trade

Foreign markets accounted for 83% of semiconductor sales by U.S.-headquartered firms in 2015, reflecting the fact that many end-user industries, such as assembly of computers and consumer electronics, are located mainly in Asia.[51] Total exports of U.S.-made semiconductors and related devices registered $41.8 billion in 2015, a reduction of 2% from the previous year. Appreciation in the value of the U.S. dollar has made American factories' goods more expensive in international markets, potentially contributing to the loss in market share.[52]

Mexico, China, Malaysia, South Korea, and Taiwan ranked as the top five U.S. export markets in 2015. These countries are large producers of consumer electronics, telecommunications equipment, and information and communications technologies, all of which rely heavily on semiconductors as a principal component. According to data from the United States International Trade Commission (USITC), in 2015, semiconductors represented the top U.S. high-tech export by value and the fourth-largest overall export by value, behind civilian aircraft, petroleum refinery products, and automobiles.[53]

Imports of semiconductors totaled $41.7 billion in 2015, expanding 3.3% from a year earlier. Malaysia, China, Taiwan, Japan, and South Korea ranked as the top five import sources for the United States. Of the top five countries from which the United States imports semiconductors, Malaysia contributed the most, accounting for more 30% of all imported semiconductors in 2015. Malaysia is an important offshore location for semiconductor packaging, assembly, and testing, including for U.S.-headquartered semiconductor firms such as Intel.[54] In 2015, 13% of U.S. semiconductor imports were from China, up from 9% in 2009.

Roughly one-third of U.S. semiconductor imports are reexported. This reflects the fact that many electronic products have complex international supply chains; thus, semiconductor products may cross several borders before being incorporated into a final product.

The semiconductor industry supports passage of the Trans-Pacific Partnership (TPP) agreement and successful conclusion of the ongoing Transatlantic Trade and Investment Partnership (TTIP) negotiations with Europe.[55] After years of negotiations, a recent success for the semiconductor industry was the expansion of the World Trade Organization (WTO) Information Technology Agreement (ITA), a plurilateral tariff-cutting agreement focused on trade in information technology goods.[56] Beginning on July 1, 2016, the expanded ITA will eliminate some tariffs immediately and phase out others by January 2024 on 201 information technology products not included in the original 1996 ITA.[57] The newly added products apply to next generation multicomponent semiconductors (MCOs), which currently face global tariff rates which generally range from 2.5% to 8.0% and can be as high as 25.0% in some countries.[58] MCOs are incorporated into a range of electronics such as smart phones, tablets, gaming consoles, e-readers, tire pressure monitors, and hand-held projectors. ITA participants committed to reconvene in 2018 to consider updating the agreement to include additional products and to possibly address non-tariff barriers in the information technology sector.[59] For additional information, see CRS Insight IN10331, *Expansion of WTO Information Technology Agreement Targets December Conclusion*, by Rachel F. Fefer.

Intellectual Property Rights

Major semiconductor producers regularly rank among the top U.S. corporate patent recipients measured by number of patents granted. In 2015, this list included Qualcomm (2,900), Intel (2,046), and Broadcom (1,086), according to data from the U.S. Patent and Trademark Office on patents granted by company.[60] The semiconductor industry supported the Defend

Trade Secrets Act (DTSA) of 2015 (P.L. 114-153) enacted on May 13, 2016. The law creates a federal private right to action for trade secret misappropriations (e.g., when an individual acquires a trade secret through improper means, including theft, bribery or espionage).[61] For more information about the protection of trade secrets, see CRS Report R43714, *Protection of Trade Secrets: Overview of Current Law and Legislation*, by Brian T. Yeh.

Congress has addressed the importation of counterfeit products, items marked or marketed as the real thing for branded versions of products, in the Foreign Counterfeit Merchandise Prevention Act (H.R. 236). The bill introduced by Representative Ted Poe would allow customs officials to share information about imported "critical" goods with intellectual property rights (IPR) holders whose copyright and trademark rights might be infringed by imports.62 In this instance, critical goods are defined as those for which counterfeits pose a danger to the health, safety, or welfare of consumers or national security, including semiconductors. Similar legislation has been introduced in previous Congresses. The semiconductor industry asserts Customs and Border Protection (CBP) has not adequately protected its IPR from growing imports of counterfeit goods, and that CBP efforts to collaborate with the private sector to identify and enforce IPR violations have been inadequate.[63]

GLOBAL COMPETITION

East Asia

American companies dominated worldwide production of semiconductors until the 1970s.[64] In the 1980s, when Japan captured the majority of the global DRAM market, the U.S. government alleged that Japanese companies achieved this position due to the Japanese government's protection of its domestic market, stifling the sale of U.S. semiconductors in Japan.[65] The U.S. government responded to this development in several ways, including seeking a bilateral agreement to open the Japanese market to U.S. semiconductors and providing federal funding for a research consortium to support U.S. technological competitiveness in the field. These efforts produced the 1986 U.S.-Japan Semiconductor Agreement and the 1987 formation of SEMATECH (short for Semiconductor Manufacturing Technology), a consortium of semiconductor companies.[66] SEMATECH and "The Japanese Challenge" are discussed later in this report.

Since the early 1990s, Japan's share of the global semiconductor market has fallen significantly. Several Japanese fabs have closed, and some producers have gone bankrupt. In 2015, only three Japanese chipmakers—Toshiba, Renesas Electronics, and Sony—were among the top 20 producers worldwide ranked by revenue.

As the market positions of Japanese companies have declined, companies based elsewhere in East Asia have become prominent global suppliers, mostly in the DRAM segment of the market. South Korea's Samsung Electronics and SK Hynix are now the second- and third-largest semiconductor companies in the world. According to data from Statistica, an industry statistics portal, at the end of 2015, Samsung held 46.4% of the global DRAM market, followed by SK Hynix at 27.9%, and Micron at 18.9%.[67] The growth of the South Korean semiconductor industry has been supported and nurtured by government funding and the financial backing of large, family-controlled industrial conglomerates known as chaebols. The chaebols play a central role in South Korea's economy.[68]

Taiwan has become the world's leading location for semiconductor foundry manufacturing. Taiwan's semiconductor foundry industry is dominated by two contract manufacturers, Taiwan Semiconductor Manufacturing Company (TSMC) and United Microelectronics Company (UMC).[69] Both TSMC and UMC were established and directly funded by the Taiwanese government in the 1980s through a variety of grants, low-interest loans, and other subsidies, although both are organized as private enterprises.[70]

China

In 2014, China accounted for close to 57% of the worldwide consumption of integrated circuits.[71] However, the country plays a limited role in the production of semiconductors. A 2014 study by the East-West Center, a nonpartisan research group established by Congress, reported that up to 80% of the semiconductors used in Chinese electronics manufacturing are imported.[72]

According to a PricewaterhouseCoopers (PWC) report on China's semiconductor industry, semiconductor manufacturers in China, including indigenous Chinese firms and multinational semiconductor firms, accounted for 13.4% of the worldwide semiconductor industry by revenue in 2014, up from 12.0% in 2013 and 11.6% in 2012.[73] In 2014, China's semiconductor industry revenues rose to $77.3 billion, up from $40.5 billion in 2013 and $34.2 billion in 2012.

Currently, non-Chinese semiconductor companies dominate the Chinese market. Dieter Ernst of the East-West Center notes, "China's domestic semiconductor manufacturing (i.e., wafer fabrication) technology and capabilities have failed to keep up with the country's IC design capabilities and needs."[74] The same report notes China's wafer fabrication plants "are using older technology and used equipment, reflecting China's focus on light-emitting diode (LED) and other applications that do not require leading-edge semiconductors."[75] Similarly, an analysis of the Chinese integrated circuit market by IBISWorld, a market research firm, found that many Chinese chips are "low-end."[76]

As shown in *Table 3*, of the 94 advanced 300mm wafer fabrication plants in operation worldwide in 2015, only nine were located in China.[77] Of these, three were owned by foreign companies: Intel, Samsung, and SK Hynix. In addition, Taiwanese semiconductor manufacturer TSMC has announced its intention to build a 300mm fab facility in China.[78]

Table 3. Worldwide 300mm Semiconductor Fab Count
Number of Operating Fabs by Country or Region

Country/Region	2011	2013	*2015*
Taiwan	21	23	28
United States	15	16	17
Japan	16	16	17
South Korea	10	10	12
China	7	7	9
Europe & Mideast	7	6	6
Southeast Asia	4	4	5
Total	80	82	*94*

Source: SEMI Worldwide Fab Forecast, April 2016. SEMI provided these statistics to CRS by email on April 26, 2016.

In June 2014, the Chinese central authorities published an ambitious plan, *Guidelines to Promote National Integrated Circuit Industry Development*, "with the goal of establishing a world-leading semiconductor industry in all areas of the integrated circuit supply chain by 2030."[79] The document includes measures to support an aggressive growth strategy, with the goal of meeting 70% of China's semiconductor demand from domestic production by 2025.[80]

To make China less dependent on imported chips, according to McKinsey & Company, the Chinese government intends to spend about $100 billion to

$150 billion on the development of its semiconductor industry.[81] Among its objectives is to turn local chip manufacturers such as Semiconductor Manufacturing International Corporation (SMIC), which now operates three 300mm fabs in China, Shanghai Huali Microelectronics Corporation (HMLC), and Wuhan Xinxin Semiconductor Manufacturing (XMC) into major global competitors.[82]

In the past, massive efforts by the Chinese government to spur national champions have failed to bring about the desired results. China faces significant barriers to advanced production in semiconductors. Export controls and other policy barriers in Taiwan, South Korea, and the United States inhibit or prohibit the transfer of the latest technologies to Chinese firms. In 2015, for example, the United States blocked the sale of a number of advanced microprocessors to China over concerns about their use in Chinese supercomputers.[83] The extreme complexity of advanced semiconductors requires a high degree of manufacturing skill, and relatively small producers may lack the economies of scale that are important to driving down unit costs.

Consistent with China's integrated circuit development plan, several Chinese companies have pursued acquisitions of foreign companies. In 2015, Tsinghua Unigroup, a Chinese state-owned enterprise (SOE), proposed to acquire Micron Technologies for $23 billion.[84] After several media outlets reported on the proposed acquisition, some Members of Congress raised concerns with Secretary of the Treasury Jacob Lew about the potential national security and economic ramifications of allowing a Chinese SOE to acquire a major U.S. technology firm, especially the principal American manufacturer of computer memory chips."[85] The acquisition was never realized. In addition, Tsinghua has sought to acquire three Taiwan-based chip packaging companies and reportedly targeted SK Hynix.[86] In 2016, state-backed Chinese investors abandoned a bid to buy one of America's oldest semiconductor manufacturers, Fairchild Semiconductor, and a unit of Tsinghua terminated a plan to buy 15% of Western Digital, which makes hard disk drives.[87]

It is not known whether any of the proposed transactions involving Chinese buyers faced objections from the Committee on Foreign Investment in the United States (CFIUS). The interagency committee reviews transactions that could result in control of a U.S. business by a foreign person to determine their potential effect on national security.

Europe

Europe's semiconductor industry includes firms such as STMicroeletronics (formed in 1986 by the merger of SGS Microelectronica of Italy and Thomson Semiconductor of France), Infineon Technologies (formed in 1999 as a spinoff from Siemens' semiconductor operations), and NXP Semiconductors (founded by Philips in 2006). These three European-headquartered firms ranked among the world's top 20 semiconductor firms by revenue in 2015.[88] Measured by advanced 300mm wafer fabrication production capacity, Europe accounted for 3.0% of worldwide production in 2015 (see *Table 1*).[89] European-headquartered semiconductor companies tend to specialize in niche markets such as semiconductors for automobiles and industrial electronics.[90] According to a 2013 communication by the European Commission, Europe made up about 50% of worldwide automotive electronics production and around 35% of global industrial electronics production.[91] The Commission also stated that Europe is strong in manufacturing electronics for energy applications, accounting for about 40% of global production in that market, and in designing electronics for mobile telecommunications. Of the $335 billion in global semiconductor sales, European-headquartered semiconductor firms accounted for about $34 billion in sales in 2015, according to figures from WSTS.[92]

In May 2013, the Commission announced an initiative to support the European semiconductor industry.[93] The initiative, set to run from 2014 to 2020, aims to increase Europe's share of global semiconductor manufacturing to at least 20% by the end of the decade by providing $11 billion (€10 billion) in public and private funding for R&D activities that it hopes will trigger about $113 billion (€100 billion) in industry investment in manufacturing. The initiative calls for a multipronged approach that includes easier access to capital financing by qualified companies; pooling EU, national, and regional subsidies to enable larger-scale projects; and, improving worker training.[94]

THE FEDERAL ROLE IN SEMICONDUCTORS

The federal government has played a major role in supporting the U.S. semiconductor industry since the late 1940s. That role, however, has changed considerably over time. In the early years, federal support for the nascent industry included research funding; support for the development of increasingly powerful computers; and, serving as an early adopter of semiconductor-enabled technologies, creating a market through defense and space-related acquisitions. From the late 1980s through the mid-1990s, the federal role centered on reversing a perceived loss of U.S. competitiveness in semiconductors through the initiation and funding of an industry research consortium. More recently, the federal role has focused on support for research to extend the life of current semiconductor technologies and to develop the scientific and technological underpinnings for revolutionary successor technologies.

Early Efforts in Computing

Two developments in the late 1940s, computers and transistors, laid the foundation for development of the semiconductor and computing industries. The first was the Electronic Numerical Integrator and Computer (ENIAC), the first general-purpose electronic digital computer, which was announced in 1946. The Army Ballistic Research Laboratory funded development of the ENIAC at the University of Pennsylvania to calculate artillery firing tables. With semiconductor devices still in the future, the ENIAC used thousands of vacuum tubes, crystal diodes, relays, resistors, and capacitors, making it large enough to fill a 30-by-50-foot room. The second major development came in 1947 when Bell Telephone Laboratories (known broadly as Bell Labs), building on federal World War II research investments, invented the transistor, a semiconductor device capable of regulating the flow of electricity.[95]

For the next decade, engineers sought to increase computer performance by overcoming the "tyranny of numbers," a term referring to the need to connect all of a computer's components to each other, a task requiring the hand-soldering of each connection. As the number of components grew to increase computing power, so did the number of connections required, adding to complexity, cost, and reliability issues. The Army Signal Corps attempted to address these challenges by funding a program which sought to make all components the same size and shape, with the wiring built in, so they could be snapped together to form a circuit without the need for soldering. A different solution was developed in 1958 by Texas Instruments with the invention of the integrated circuit (IC), which incorporated resistors, capacitors, and transistors on a single sliver of the semiconducting element germanium. Shortly thereafter, Fairchild Semiconductor developed a silicon-based IC that included a final layer of metal, parts of which could be removed to create the necessary connections, making it more suitable for mass production.[96] While the invention of the IC was accomplished without direct federal funding, government purchases of ICs for military, space, and other uses supplied the initial demand that allowed manufacturers to reduce costs. As late as 1962, government purchases accounted for 100% of total U.S. IC sales.[97]

The Japanese Challenge

Throughout the 1960s and 1970s, the U.S. semiconductor industry grew rapidly and was largely unchallenged on the world stage. While the U.S. share of global semiconductor *consumption* fell from an estimated 81% in 1960 to around 57% in 1972, the U.S. share of global *production* remained at around 60%.[98] However, the rapid ascent of Japan's semiconductor industry in the early 1980s stirred concerns about the potential decline in the competitive position of the U.S. semiconductor industry. By the late 1980s, the U.S. share of global semiconductor sales fell below 40%.[99]

In 1987, the Defense Science Board's Task Force on Semiconductor Dependency found that U.S. leadership in semiconductor manufacturing was rapidly eroding and that not only was "the manufacturing capacity of the U.S. semiconductor industry...being lost to foreign competitors, principally Japan... but of even greater long-term concern, that technological leadership is also being lost." In addition to the decline in the semiconductor device industry, the task force found that "related upstream industries, such as those that supply silicon materials or processing equipment, are losing the commercial and technical leadership they have historically held in important aspects of process technology and manufacturing, as well as product design and innovation."[100]

The task force recommended the formation of an industry-government consortium to "develop, demonstrate and advance the technology base for efficient, high yield manufacture of advanced semiconductor devices." Describing this as the "principal and most crucial recommendation of the Task Force," the report estimated that "the initial capitalization of the Institute by its industrial members would be on the order of $250 million," and recommended federal support of approximately $200 million per year for five years through the Department of Defense.[101]

In 1987, 14 U.S. semiconductor firms founded the SEMATECH (short for Semiconductor Manufacturing Technology) research consortium in Austin, TX. From FY1988 to FY1996, Congress provided a total of approximately

$870 million to SEMATECH through the Defense Advanced Research Projects Agency (DARPA), generally matched by contributions from the industry participants.[102]

By 1994 the U.S. semiconductor industry share of the global market had begun to grow again. According to the National Academy of Sciences, "SEMATECH was widely perceived by industry to have had a significant impact on U.S. semiconductor manufacturing performance in the 1990s."[103] A 1992 evaluation by the Government Accounting Office of the federal partnership in SEMATECH's found that

> SEMATECH has shown that a government-industry R&D consortium can help improve a U.S. industry's technological position by developing advanced manufacturing technology. Whether this can be replicated and what conditions would lead to this result in other cases is uncertain.[104]

Among SEMATECH's leading detractors was Cypress Semiconductor chief executive officer T.J. Rodgers. In a 1998 paper, Rodgers asserted that SEMATECH's federal funding was a subsidy to large, wealthy companies; that hundreds of smaller semiconductor firms were excluded from participating in SEMATECH due to its minimum $1 million annual dues; and that SEMATECH engaged in "hold back" contracts that denied non-SEMATECH firms access to technology that emerged from SEMATECH research. Summing up, Rodgers stated that SEMATECH "used the combined resources of its members and the government to create a competitive advantage, and it kept its secrets from its competitors."[105]

In July 1994, the SEMATECH Board of Directors voted to not accept any additional federal funding after FY1996. The consortium continued to operate on industry funding, allowing foreign-based companies to join. Following the departure of members Intel and Samsung in 2015, SEMATECH was absorbed by the State University of New York Polytechnic Institute and is now based in Albany, NY.

Current Federal Efforts

The federal government has continued to support a wide array of semiconductor research activities.

A major area of research has focused on a successor to complementary metal–oxide– semiconductor (CMOS) technology, which has been the basis of semiconductor manufacturing for half a century. Research and development leading to a continual reduction in the size of components on each chip has enabled CMOS-based semiconductors to become more powerful, more energy-efficient, and less expensive. However, it is widely believed that "as the dimensions of critical elements of devices approach atomic size, quantum tunneling and other quantum effects [will] degrade and ultimately prohibit further miniaturization of conventional devices."[106] This has spurred additional federal efforts to develop other semiconductor technologies.

In July 2015, President Obama issued an executive order establishing the National Strategic Computing Initiative (NSCI) "to create a cohesive, multi-agency strategic vision and federal investment strategy, executed in collaboration with industry and academia, to maximize the benefits of HPC [high performance computing] for the United States." A key objective of the NSCI is to establish, "over the next 15 years, a viable path forward for future HPC systems even after the limits of current semiconductor technology are reached."[107] The executive order designates the U.S. Department of Energy (DOE), the National Science Foundation (NSF), and the Department of Defense (DOD) as the lead agencies, and designates the Intelligence Advanced Research Projects Activity and National Institute of Standards and Technology (NIST) as foundational research and development agencies.

Other federal efforts include the following:

- Semiconductor Technology Advanced Research Network (STARnet). STARnet, a partnership between DARPA and semiconductor and defense companies, is a collaborative network of research centers focused on "finding paths around the fundamental physical limits threatening the long-term growth of the microelectronics industry."[108]
- Secure and Trustworthy Cyberspace: Secure, Trustworthy, Assured and Resilient Semiconductors and Systems (SaTC: STARSS). SaTC: STARSS is a joint research effort of NSF and the Semiconductor Research Corporation (SRC) [109] focused on new strategies for semiconductor architecture, specification and verification, to increase resistance and resilience to tampering and to improve authentication throughout the supply chain.[110]
- Nanoelectronics for 2020 and Beyond. Nanoelectronics for 2020 and Beyond is an effort organized under the National Nanotechnology Initiative (NNI)[111] "to discover and use novel nanoscale fabrication

processes and innovative concepts to produce revolutionary materials, devices, systems, and architectures."[112] Congress has provided approximately $530 million for the Nanoelectronics for 2020 and Beyond initiative since FY2011, primarily through NSF, DOD, and NIST. Specific projects include the Nanoelectronics Research Initiative, a public-private partnership with the SRC and STARnet.

- Energy-Efficient Computing: from Devices to Architectures (E2CDA). E2CDA is a joint initiative between NSF and SRC focused on the development of technologies to reduce the amount of energy it takes to manipulate, store, and transport data.

NATIONAL SECURITY CONCERNS

For decades, many have argued that maintaining a domestic manufacturing capability for the most advanced semiconductor products is necessary for national security. Proponents of this view claim dependence by the U.S. military on foreign suppliers of semiconductors, especially those that are hostile or may become hostile to U.S. interests, is not acceptable due to the military's reliance on semiconductors as a vital and ubiquitous component in U.S. weapons and defense systems. However, the high costs of maintaining a domestic semiconductor production capability for critical military inputs may result in more expensive weapons systems.[113]

In 2003, then-Deputy Secretary of Defense Paul Wolfowitz wrote in an unclassified memo the "country needs a defense industrial base that includes leading edge, trusted commercial suppliers for critical integrated circuits used in sensitive defense weapons, intelligence, and communications systems."[114] As a follow-up to the memo, the Department of Defense (DOD) implemented a trusted supplier program (originally named the trusted foundry program) in 2004, whereby the government pays a fee to U.S. companies to guarantee the access and reliability of components that are important to national defense.[115]

Under the program, IBM's fabrication facilities supplied advanced semiconductors to DOD as the sole source contractor. In 2014, however, IBM announced that the United Arab Emirates-owned GlobalFoundries would acquire its unprofitable microelectronics fabrication facilities in Vermont and New York.[116] The Committee on Foreign Investment in the United States (CFIUS)[117] reviewed the transaction and in July 2015 said that it would not prohibit the acquisition.[118] GlobalFoundries also obtained the appropriate accreditations to be a DOD trusted supplier. According to recent news reports, in June 2016, DOD reached a seven-year agreement with GlobalFoundries to supply microchips until 2023.[119]

In October 2015, the House Armed Services Subcommittee on Oversight and Investigations held a hearing that considered the long-term viability of the DOD trusted supplier program in light of the shrinking number of domestic microelectronics manufacturers and other ways the semiconductor industry has changed.[120] Future policy options under consideration include identifying additional U.S.-based trusted foundries with leading-edge manufacturing capability; exploring alternative manufacturing approaches, which may incorporate non-U.S. made semiconductor parts; or, establishing a government-owned fabrication facility.[121] Beyond manufacturing, the trusted supplier program also includes firms that provide other services in the semiconductor supply chain, including design, assembly, and testing.[122]

Despite national security concerns, DOD is heavily reliant on the commercial supply chain, which includes many non-U.S. suppliers, for most of its electronic hardware and the trusted supplier program is used for only a small fraction of the chips in defense systems. In summer 2015, DOD's Office of Manufacturing and Industrial Base Policy began a study on the microelectronics industrial base, which when finished is expected to include recommendations on strategies to increase DOD's access to trusted microelectronics manufacturers.[123]

End Notes

[1] Semiconductor Industry Association (SIA), "Global Semiconductor Sales Top $335 Billion in 2015," press release, February 1, 2016.

[2] Audi, "Immense Significance for Innovations: Semiconductors," http://www.audi.com/com/brand/en/vorsprung_durch_technik/content/2014/10/underrated-qualities-semiconductors.html.

[3] Robert N. Charette, "This Car Runs on Code," IEEE Spectrum, February 1, 2009, http://spectrum.ieee.org/transportation/systems/this-car-runs-on-code.

[4] Dylan Tweney, "April 19, 1965: How Do You Like It? Moore, Moore, Moore," Wired, April 19, 2010, http://www.wired.com/2010/04/0419moores-law-published/; Stephen Shankland, "Moore's Law: The Rule that Really Matters in Tech," CNet, October 15, 2012, http://www.cnet.com/news/moores-law-the-rule-that-really-matters-in-tech/.

[5] Some features of chips are now under 10 nanometers (nm), and Intel anticipates a 5nm process in 2019. For comparison, eight hydrogen atoms side-by-side would measure about one nanometer. This size-scale presents challenges for manufacturability and adverse effects related to heat and unique quantum phenomena. Each reduction in feature size is considered a move to a new generation of manufacturing technology, and each new generation

generally represents a doubling of the density of transistors on a silicon wafer, creating ever more powerful semiconductors.

[6] For differing opinions on the future prospects of silicon-based semiconductors, see "Double, Double, Toil and Trouble," The Economist, March 12, 2016, and Bret Swanson, Moore's Law at 50: The Performance and Prospects of the Exponential Economy, American Enterprise Institute, November 2015, pp. 14-15.

[7] Conceptually, quantum computing relies on quantum phenomena to expand the number of states in which data can be encoded and stored; optical computing relies on light, rather than electric current, to perform calculations; and, neuromorphic computing relies on mimicking the architecture and processing used by biological nervous systems.

[8] Angelo Zino, Semiconductors & Semiconductor Equipment, S&P Capital IQ, May 2016, pp. 19-24.

[9] SIA, Factbook 2016, April 1, 2016, p. 2.

[10] IRS Technology, "Global Semiconductor Market Slumps in 2015, IRS Says," press release, April 4, 2016, https://technology.ihs.com/576301/global-semiconductor-market-slumps-in-2015-ihs-says.

[11] Fabrication is the multi-step process used to create integrated circuits, including microprocessors, memory, and microcontrollers. The entire manufacturing process takes six to eight weeks and is performed in fabs that require clean rooms. Integrated device manufacturers (IDM) can also provide their chip manufacturing capacity to companies that do not have their own fabrication facilities. In some instances, IDMs lack sufficient capacity and outsource some of their chip manufacturing to contract foundries.

[12] Ulrich Naeher, Sakae Suzuki, and Bill Wiseman, The Evolution of Business Models in a Disrupted Value Chain, McKinsey & Company, McKinsey on Semiconductors, November 1, 2011, pp. 33-34, http://www.mckinsey.com/ industries/semiconductors/our-insights.

[13] World Semiconductor Trade Statistics (WSTS), "Worldwide Semiconductor Market is Expected to be Slightly Positive in 2016 and Grow Moderately in 2017," press release, February 25, 2016, https://www.wsts.org/PRESS/ Recent-News-Release.

[14] SIA, Factbook 2016, April 1, 2016, p. 2.

[15] SIA, "Global Semiconductor Sales Top $335 Billion in 2015," press release, February 1, 2016.

[16] WSTS, "Worldwide Semiconductor Market is Expected to be Slightly Positive in 2016 and Grow Moderately in 2017," press release, February 25, 2016.

[17] SIA and Nathan Associates, Beyond Borders: The Global Semiconductor Value Chain, May 2017, pp. 41-42.

[18] Angelo Zino, Industry Surveys Semiconductors & Semiconductor Equipment, S&P Capital IQ, May 2016, p. 50.

[19] SIA and Nathan Associates, Beyond Borders: The Global Semiconductor Value Chain, May 2017, p. 41.

[20] Anne Shields, Why DRAM Pricing is a Key Concern for Micron, Market Realist, July 10, 2015, http://marketrealist.com/2015/07/dram-pricing-key-concern-micron/.

[21] Jeho Lee, "The Chicken Game and the Amplified Semiconductor Cycle: The Evolution of the DRAM Industry from 2006 to 2014," Seoul Journal of Business, vol. 21, no. 1 (June 2015), pp. 2 and 22.

[22] Intel, "Intel Announces Restructuring Initiative to Accelerate Transformation," press release, April 19, 2016, http://files.shareholder.com/downloads/INTC/1958830663x0x8866 62/6D73A0D5-A8CD-48A2-96E7-5234880B6304/Press_Release_Q1_2016_restructuring_-_FINAL.pdf.

[23] USITC, The Information Technology Agreement, Advice and Information on the Proposed Expansion: Part 2, February 2013, pp. 3-9.
[24] Kokomo Operations Overview, http://www.slideshare.net/boilerfunk/kokomo-semiconductors-introduction-sep-3- 2010. Kokomo's website notes that through the years the company has broadened its customer base to include other automotive component suppliers, personal computer manufacturers, and avionics electronics suppliers.
[25] First Research, Semiconductor & Other Electronic Component Manufacturing, March 21, 2016.
[26] The front-end manufacturing process covers everything from the creation of the silicon wafer to the production of integrated circuits on the wafer, and includes lithography, deposition, etching and striping, inspection and doping.
[27] Falan Yinug, Challenges to Foreign Investment in High-Tech Semiconductor Production in China, USITC, May 2009, p. 16, https://www.usitc.gov/publications/332/journals/semiconductor_production.pdf.
[28] The year-end forecast for sales of semiconductor manufacturing equipment in North America was $5.6 billion in 2015. SEMI, "Semiconductor Equipment Sales Forecast," December 15, 2015.
[29] U.S. Government Accountability Office (GAO), Export Controls: Challenges with Commerce's Validated End-User Program May Limit Its Ability to Ensure that Semiconductor Equipment Exported to China is Used as Intended, GAO-08-1094, September 2008, p. 1, http://www.gao.gov/assets/290/282096.pdf.
[30] U.S. Census Bureau, Statistics of U.S. Businesses, 2013, http://www.census.gov/econ/susb/.
[31] An industry's value added measures its contribution to the economy. Industry value added based on NAICS 334413 from the U.S. Census Bureau's Annual Survey of Manufacturers.
[32] Semiconductor price data based on NAICS 334413 from the U.S. Bureau of Labor Statistics (BLS) producer price index program, http://www.bls.gov/ppi/.
[33] SIA and Nathan Associates, Beyond Borders: The Global Semiconductor Value Chain, May 2017, p. 5.
[34] National Science Foundation (NSF), Domestic R&D Paid by the Company and Performed by the Company and Others as a Percentage of Domestic Net Sales, by Industry and Company Size: 2012, Table 19, Business Research and Development Innovation: 2012, October 29, 2015, http://www.nsf.gov/statistics/2016/nsf16301/pdf/tab19.pdf.
[35] SIA, The U.S. Semiconductor Industry, 2016 Factbook, April 1, 2016, pp. 18-19.
[36] The permanent tax credit was included in the Consolidated Appropriations Act, 2016 (P.L. 114-113), enacted December 18, 2015. The R&D tax credit has expired 17 times since it was first established in 1981. For more information see, CRS Report RL31181, Research Tax Credit: Current Law and Policy Issues for the 114th Congress, by Gary Guenther.
[37] SIA, "Semiconductor Industry Hails Passage of Permanent R&D Credit," press release, December 18, 2015, http://www.semiconductors.org/news/2015/12/18/press_releases_2015/semiconductor_industry_hails_passage_of_permanent_r_d_credit/.
[38] BLS, Quarterly Census of Employment and Wages (QCEW) for NAICS 334413, http://www.bls.gov/cew/.
[39] Average wage data are from BLS's QCEW program. 2015 data are preliminary.
[40] SIA, U.S. Semiconductor Industry Employment, April 2013, p. 3, http://www.semiconductors.org/clientuploads/Jobs%20Rollout/Jobs%20Issue%20Paper_April_2013.pdf.
[41] State employment data are from the BLS QCEW program.
[42] Angelo Zino, Semiconductors & Semiconductor Equipment: Industry Surveys, S&P Capital IQ, May 2016, p. 47; Jim Handy, "Why Are Computer Chips So Expensive? Why Are

Computer Chips So Expensive?," Forbes, April 30, 2014. The industry anticipates a shift to 450mm wafer production, which could reduce costs by 30% and allow for greater throughput. Most industry executives expect this shift after 2020. See, KPMG, Global Semiconductor Outlook 2016: Seismic Shifts Underway, p.20, http://www.kpmg.com/US/en/industry/technology/Documents/kpmg-globalsemiconductor-outlook-2016.pdf.

[43] Capital expenditures based on NAICS 3344 (semiconductors and other electronic component manufacturing) from the U.S. Census Bureau's Annual Capital Expenditures Survey, http://www.census.gov/programs-surveys/aces.html.

[44] IC Insights, S. Korean and Taiwanese Companies Control 56% of Global 300mm Fab Capacity, December 18, 2014, p. 2, http://www.icinsights.com/data/articles/documents/742.pdf.

[45] Charles Wessner and Alan Wolff, Rising to the Challenge: U.S. Innovation for the Global Economy, National Research Council, 2012, p. 340.

[46] Robert C. Leachman and Chien H. Leachman, "Globalization of Semiconductors: Do Real Men Have Fabs, or Virtual Fabs?" in Martin Kenney with Richard Florida, eds., Locating Global Advantage: Industry Dynamics in the International Economy. Stanford, CA: Stanford University Press, November 18, 2003, p. 226.

[47] Intel, 2015 Annual 10-K report, p.10.

[48] Micron, 10-K Annual Report, October 27, 2015, p. 20, http://investors.micron.com/. All other domestic producers have either shut down or outsourced their DRAM manufacturing to foundries abroad.

[49] Texas Instruments, 2014 Corporate Citizenship Report, May 21, 2015, p. 4.

[50] SEMI provided data to CRS from its proprietary Fab Construction Monitor database. The new fab construction totals include two small LED fabs in the United States.

[51] Falan Yinug, Made in America: The Facts about Semiconductor Manufacturing, SIA, August 2015, p. 4. http://blog.semiconductors.org/blog/sia-white-paper-facts-about-semiconductor-manufacturing; SIA, The Semiconductor Industry Association's Comments to the President's Economic Recovery Advisory Board's Tax Reform Subcommittee, October 15, 2009, p. 2, https://www.whitehouse.gov/assets/formsubmissions/109/df34744d4311400 7a53ecfb8479b7898.pdf.

[52] Darryle Ulama, Earth Potential: International Competition May Outpace Growth Despite Increased Demand, IBISWorld, Report 33441A: Semiconductor & Circuit Manufacturing in the U.S., December 2015, p. 10.

[53] CRS analysis of U.S. trade data by six-digit NAICS code from the USITC's dataweb.

[54] Luis Abad, Ngozika Amalu, and Ramona Lohan, et al., The Malaysian Semiconductor Cluster, Microeconomics of Competitiveness, May 8, 2015, p. 8, http://www.isc.hbs.edu/resources/courses/moc-course-at-harvard/Documents/pdf/student-projects/Malaysia_Semiconductor_Cluster_2015.pdf.

[55] SIA, 2016 Policy Priorities, http://www.semiconductors.org/clientuploads/Resources/SIA%202016%20Policy%20Priorities%201%20Pager%20FINAL.pdf.

[56] World Trade Organization (WTO), WTO Members Conclude Landmark $1.3 Trillion Trade Deal, December 16, 2015, https://www.wto.org/english/news_e/ news15_e/ita_16dec15_e.htm. (A plurilateral agreement involves a subset of countries that often negotiate to liberalize trade in a specific sector.)

[57] The original ITA signed in 1996 was intended to cover all semiconductors and integrated circuits under semiconductor harmonized tariff scheduled (HTS) headings 8541 and 8542, meaning these products entered most markets duty-free. However, the 1996 ITA did not include a mechanism to cover new advanced technology products.

[58] Devi Keller, The Benefits of Including Multi-Component Semiconductors in an Expanded Information Technology Agreement, SIA, December 8, 2014, http://blog.semiconductors.org/blog/the-benefits-of-including-mcos-in-anexpanded-information-technology-agreement.

[59] WTO, "Chinese Taipei, Thailand Confirm Acceptance of Landmark IT Deal," press release, July 28, 2015, https://www.wto.org/english/news_e/news15_e/ita_28jul15_e.htm.

[60] U.S. Patent and Trademark Office, All Technologies Report, January 1, 1991—December 31, 2015, March 2016, http://www.uspto.gov/web/offices/ac/ido/oeip/taf/all_tech.pdf.

[61] SIA, "Semiconductor Industry Commends Passage of Legislation to Protect Trade Secrets," press release, April 27, 2016, http://www.semiconductors.org/news/2016/04/27/press_releases_2015/semiconductor_industry_commends_passage_ofjegislation_to_protect_trade_secrets/.

[62] SIA, Winning the Battle Against Counterfeit Semiconductor Products, August 2013, http://www.semiconductors.org/clientuploads/Anti-Counterfeiting/SIA%20Anti-Counterfeiting%20Whitepaper.pdf.

[63] U.S. Congress, House Committee on the Judiciary, Subcommittee on Crime, Terrorism, and Homeland Security, Hearing on H.R. 4223, the Safe Doses Act"; H.R. 3668, the Counterfeit Drug Penalty Enhancement Act of 2011; and H.R. 4216, the "Foreign Counterfeit Prevention Act", 112th Cong., 2nd sess, March 28, 2012, Testimony of Mr. Travis D. Johnson.

[64] National Defense University, Industry Study: Electronics, Spring 2015, p. 1.

[65] Douglas A. Irwin, The Political Economy of Trade Protection, National Bureau of Economic Research, The U.S.- Japan Semiconductor Trade Conflict, January 1996, p. 7. A version of the chapter is available at http://www.nber.org/ chapters/c8717.pdf.

[66] The 1986 U.S.-Japan Semiconductor Agreement included three major provisions: (1) Japan agreed to open its markets to U.S. semiconductors; (2) Japan committed to the goal of a 20% foreign share of the Japanese market by 1992 (which was not reached during the life of the agreement); and, (3) Japan agreed to stop dumping in third markets.

[67] Statistica, The Statistics Portal, "Global Market Share of the DRAM Chip Market from 1st Quarter 2011 to 1st Quarter 2016 by Vendor," http://www.statista.com/statistics/271726/global-market-share-held-by-dram-chip-vendors since-2010/.

[68] Nayanee Gupta, David W. Healey, and Aliza M. Stein, et al., Innovation Policies of South Korea, Institute for Defense Analyses, August 2013, https://www.ida.org/~/media/Corporate/Files/Publications/STPIPubs/ida-d-4984.ashx.

[69] Taiwan Semiconductor Industry Association, Overview of Taiwan Semiconductor Industry, 2015, p. 9, http://www.tsia.org.tw/Uploads/2015%20Overview-Final.pdf.

[70] Tain-Jy Chen, Taiwan's Industrial Policy Since 1990, Department of Economics, National Taiwan University, April 2014, p. 9.

[71] PWC, China's Impact on the Semiconductor Industry: 2015 Update, October 2015, p. 5, https://www.pwc.com/gx/ en/technology/pdf/china-semicon-2015-report1-3.pdf.

[72] Dieter Ernst, From Catching Up to Forging Ahead? China's Prospects in Semiconductors, East-West Center, November 2014, p. 7, http://www.eastwestcenter.org/system/tdf/private/ernst-semiconductors2015_0.pdf?file=1& type=node&id=35320.

[73] PricewaterhouseCoopers (PWC), A Decade of Unprecedented Growth: China's Impact on the Semiconductor Industry 2014 Update, January 2015, p. 18, http://www.pwc.com/gx/en/technology/chinas-impact-on-semiconductorindustry/assets/china-semicon-2014.pdf.

[74] Dieter Ernst, From Catching Up to Forging Ahead: China's Policies for Semiconductors, East-West Center, 2015, p. 7, http://www.eastwestcenter.org/system/tdf/private/ernst-semiconductors2015_0.pdf?file=1&type=node&id=35320.

[75] Ibid., p. 10.
[76] IBISWorld, Integrated Circuit Manufacturing in China, April 2016, p. 6.
[77] PWC, China's Impact on the Semiconductor Industry: 2015 Update, Technology Institute Full Report, March 2016, p. 24, http://www.pwc.com/gx/en/technology/pdf/china-semicon-2015-report-1-5.pdf.
[78] Alan Patterson, "TSMC Aims to Build Its First 12-inch Fab in China," EE Times, December 17, 2015.
[79] International Trade Administration (ITA), 2015 Top Markets Report: Semiconductors and Semiconductor Manufacturing Equipment, A Market Assessment Tool for U.S. Exporters, July 2015, p. 13.
[80] "Chips on their Shoulders," The Economist, January 23, 2016.
[81] Christopher Thomas, A New World Under Construction: China and Semiconductors, McKinsey & Company/, November 2015 http://www.mckinsey.com/global-themes/asia-pacific/a-new-world-under-construction-china-andsemiconductors.
[82] SMIC, SMIC Presentation, March 2016, p. 10, http://www.smics.com/download/ir_presentation.pdf.
[83] Department of Commerce, Bureau of Industry and Security, "Addition of Certain Persons to the Entity List; and Removal of Person from the Entity List Based on a Removal Request," 80 Federal Register 8524-8529, February 18, 2015.
[84] Allison Gatlin, "Micron Snubs Tsinghua, Favors Another Chinese Partnership: Analyst," Investor's Business Daily, February 16, 2016, http://www.investors.com/news/technology/micron-snubs-tsinghua-favoring-another-chinesepartnership-analyst/.
[85] "CFIUS Likely to Investigate, Require Changes if Chinese SOE Vies for Micron," Inside U.S. Trade, August 28, 2015.
[86] "SK-Hynix Says Reject Tsinghua Unigroup Collaboration Offer," Reuters, November 26, 2015.
[87] James Fontanella-Khan, "Fairchild rejects $2.6bn Chinese offer," Financial Times, February 16, 2016. Joshua Jamerson and Eva Dou, "Chinese Firm Ends Investment in Western Digital, Complicating SanDisk Tie-Up," Wall Street Journal, February 23, 2016.
[88] IHS Global Insight, "Preliminary 2015 Semiconductor Market Shares," press release, December 8, 2015, https://technology.ihs.com/553230/preliminary-2015-semiconductor-market-shares.
[89] IC Insights, Global Wafer Capacity 2016-2020, IC Insights, http://www.icinsights.com/services/global-wafercapacity/report-contents/.
[90] Page Tanner, Germany to Drive Growth in European Semiconductor Market, Market Realist, December 24, 2015, http://marketrealist.com/2015/12/germany-drive-growth-european-semiconductor-industry/.
[91] European Commission, A European Strategy for Micro- and Nanoelectronic Components and Systems, May 23, 2013, p. 6, https://ec.europa.eu/digital-single-market/en/news/communication-european-strategy-micro-andnanoelectronic-components-and-systems.
[92] WSTS, "Worldwide Forecasts the Semiconductor Market to Have Slight Growth by 2018," press release, June 7, 2016, https://www.wsts.org/PRESS/Recent-News-Release.
[93] European Commission, "Electronics Industry Submits Plan to Make Europe a Global Leader in Micro and Nano-Electronics," press release, February 14, 2014, http://europa.eu/rapid/press-release_IP-14-148_en.htm.
[94] The initiative was named 10/100/20 from its three main goals. SEMI, Supporting Competitive Semiconductor Advanced Manufacturing, February 24, 2014, http://www.semi.org/eu/sites/semi.org/files/docs/SEMI%20Europe%20News-Feb%2024%202014.pdf.

[95] Executive Office of the President, National Science and Technology Council, Technology in the National Interest, 1996.
[96] Nobelprize.org: The Official Site of the Nobel Prize, "The History of the Integrated Circuit," http://www.nobelprize.org/educational/physics/integrated_circuit/history.
[97] Mowery, Federal policy and the development of semiconductors, computer hardware, and computer software, Table 1.
[98] Consumption as measured in value. William F. Finan, The International Transfer of Semiconductor Technology Through U.S.-Based Firms, National Bureau of Economic Research, Working Paper No. 118, New York, NY, December 1975, http://www.nber.org/papers/w0118.pdf. Peter R. Morris, A History of the World Semiconductor Industry, The Institution of Engineering and Technology (1989), p. 141.
[99] National Research Council, Committee on Comparative National Innovation Policies: Best Practice for the 21st Century, Rising to the Challenge: U.S. Innovation Policy for the Global Economy, 2012, http://www.ncbi.nlm.nih.gov/books/NBK100307.
[100] Department of Defense, Defense Science Board, Task Force on Semiconductor Dependency, Report of Defense Science Board Task Force on Semiconductor Dependency, February 1987, http://www.dtic.mil/cgi-bin/GetTRDoc?Location=U2&doc=GetTRDoc.pdf&AD=ADA178284.
[101] Ibid.
[102] Congressional Research Service, SEMATECH: Issues and Options (IB93024), June 12, 1996, by Glenn J. McLoughlin. Report available from the author upon request.
[103] National Research Council, Policy and Global Affairs, Board on Science, Technology, and Economic Policy, Committee on Comparative Innovation Policy: Best Practice for the 21st Century, 21st Century Innovation Systems for Japan and the United States: Lessons from a Decade of Change: Report of a Symposium, 2009, p. 8, http://www.nap.edu/download/12194.
[104] U.S. Government Accountability Office, Federal Research: Lessons Learned from SEMATECH, "Highlights," RCED-92-283, September 28, 1992, http://www.gao.gov/products/RCED-92-283.
[105] Cato Institute, "T.J. Rodgers, Silicon Valley Versus Corporate Welfare, Cato Institute Brief Papers, Briefing Paper No. 37, April 27, 1998, http://www.cato.org/pubs/briefs/bp-37.html.
[106] National Science and Technology Council (NSTC), Subcommittee on Nanoscale Science, Engineering, and Technology (NSET), The National Nanotechnology Initiative: Supplement to the President's FY2017 Budget, p. 17, http://www.nano.gov/sites/default/files/pub_resource/nni_fy17_budget_supplement.pdf.
[107] Executive Order 13702, "Creating a National Strategic Computing Initiative," 80 Federal Register 46177-46180, July 29, 2015.
[108] Semiconductor Research Corporation, website, "STARnet Research," https://www.src.org/program/starnet.
[109] SRC is a U.S. non-profit research consortium established by semiconductor companies in 1982 to "define relevant research directions, explore potentially important new technologies (and transfer results to industry), [and] generate a pool experienced faculty and relevantly educated students." (Source: Semiconductor Research Corporation, SRC: Celebrating 30 Years, https://www.src.org/src/story/src-celebrating-30-years-expanded.pdf.
[110] NSF, Secure and Trustworthy Cyberspace: Secure, Trustworthy, Assured and Resilient Semiconductors and Systems Program Solicitation (NSF 14-528), http://www.nsf.gov/pubs/2014/nsf14528/nsf14528.htm#pgm_desc_txt.

[111] An NNI signature initiative is a mechanism for combining the expertise, capabilities, and resources of federal agencies to accelerate research, development, or insertion, and overcome challenges to the application of nanotechnology-enabled products.

[112] NSTC, NSET, The National Nanotechnology Initiative: Supplement to the President's FY2017 Budget, p. 17, http://www.nano.gov/sites/default/files/pub_resource/nni_fy17_budget_supplement.pdf.

[113] Daniel M. Marrujo, Trusted Foundry Program, Defense Microelectronics Activity, October 31, 2012, pp. 11-12.

[114] Department of Defense, Defense Science Board Task Force on High Performance Microchip Supply, December 2005, pp. 87-88.

[115] The trusted supplier program is jointly managed by DOD and the National Security Agency. DOD's Defense Microelectronics Activity (DMEA) certifies and accredits firms as trusted suppliers in the areas of state-of-the-art microelectronics design and manufacturing and other capabilities when they are custom-designed, custom-manufactured, or tailored for a specific DOD military end use. For a list of the more than 70 DMEA- trusted suppliers, see http://www.dmea.osd.mil/trustedic.html.

[116] GlobalFoundries is owned by the government of Abu Dhabi in the United Arab Emirates. IBM, "GlobalFoundries to Acquire IBM's Microelectronics Business," press release, October 20, 2014, https://www-03.ibm.com/press/us/en/ pressrelease/45110.wss.

[117] CFIUS is authorized to conduct national security reviews of foreign acquisitions of U.S.-based firms under section 721 of the Defense Production Act of 1950. The President has the authority to suspend or block foreign mergers and acquisitions involving U.S.-based firms if they present credible threats to national security, which includes the loss of reliable suppliers of defense-related goods and services. The CFIUS process is legally bound by strict confidentiality requirements, and it does not disclose whether a notice has been filed or the results of any filing. However, it does provide a confidential report to Congress upon the conclusion of its review.

[118] GlobalFoundries, "GlobalFoundries Obtains U.S. Government Clearance for IBM Microelectronics Business Acquisition," press release, June 29, 2015, http://www.globalfoundries.com/newsroom/press-releases/2015/06/29/globalfoundries-obtains-u.s.-government-clearance-for-ibm-microelectronics-business-acquisition.

[119] Doug Cameron, "Pentagon Hires Foreign Chips Supplier," Wall Street Journal, June 5, 2016.

[120] U.S. Congress, House Committee on Armed Services, Subcommittee on Oversight and Investigations, Assessing DOD's Assured Access to Micro-Electronics in Support of U.S. National Security Requirements, 114th Cong., 2nd sess., October 28, 2015.

[121] General Accountability Office, Trusted Defense Microelectronics: Future Access and Capabilities are Uncertain, GAO-16-185%, October 2015, pp. 4-5.

[122] The commercial semiconductor packaging and assembly industry is located mainly in Asia.

[123] For additional background, see Manufacturing and Industrial Base Policy (MIBP), http://www.acq.osd.mil/mibp/.

INDEX

A

access, 5, 7, 8, 9, 11, 12, 13, 15, 16, 19, 21, 22, 23, 25, 28, 29, 30, 31, 33, 34, 35, 36, 37, 38, 40, 43, 45, 54, 57, 60, 62, 86, 91, 103, 106, 108, 109
accounting, 91, 93, 95, 98, 103
acquisitions, 85, 102, 104, 116
adverse effects, 109
agencies, vii, 1, 3, 17, 42, 44, 49, 70, 79, 82, 107, 116
Argentina, 52, 84
Asia, 2, 36, 38, 52, 54, 67, 84, 86, 97, 99, 100, 101, 116
Asia Pacific Economic Cooperation, 52
assets, 20, 21, 47, 61, 75, 79, 111, 112, 113
atoms, 109
authentication, 36, 107
authority(ies), 22, 24, 42, 82, 101, 116
automation, 7, 92
automobiles, 88, 97, 103

B

Bahrain, 51
balance of payments, 58
banking, 3, 7, 28, 56, 59
banks, 17, 48, 79
barriers, vii, 2, 3, 7, 11, 12, 13, 14, 15, 16, 19, 21, 31, 32, 35, 36, 37, 39, 43, 46, 62, 86, 98, 102
barriers to entry, 86
base, 12, 17, 20, 28, 33, 89, 96, 102, 104, 105, 108, 109, 110, 111
BEA, 48, 49, 55, 57, 58, 63, 67, 70, 72, 73, 74, 80, 82, 83
Beijing, 48
Belgium, 67, 84
benefits, 7, 8, 9, 17, 21, 31, 33, 107, 113
bilateral, 2, 5, 28, 30, 35, 43, 99
bounds, 46, 66, 73
Brazil, 14, 19, 38, 52, 68, 69, 84
broadband, 5, 12, 22
Bureau of Labor Statistics, 94, 111
business model, 2, 9
businesses, 4, 7, 11, 12, 15, 19, 20, 21, 27, 56, 61, 62
buyer, 54, 56, 58, 61, 92, 102

C

capital intensive, 95
category a, 56, 57, 83, 84
category b, 73, 91
Census, 47, 65, 70, 73, 83, 95, 111, 112
chaebols, 100
challenges, 2, 9, 20, 23, 26, 30, 41, 42, 53, 55, 104, 109, 116

Index

China, 6, 13, 16, 19, 20, 21, 22, 23, 26, 27, 28, 29, 38, 41, 43, 48, 49, 50, 52, 68, 84, 86, 89, 92, 95, 97, 100, 101, 102, 111, 113, 114
Chinese firms, 29, 100, 102
Chinese government, 26, 27, 28, 86, 87, 101, 102
citizens, 22, 25
civil action, 33
civil society, 21
classification, 64, 83
collaboration, 7, 22, 107
commerce, 3, 7, 8, 11, 12, 15, 18, 23, 25, 26, 30, 31, 32, 33, 35, 36, 37, 38, 39, 40, 42, 46, 50, 61, 64, 65, 73, 83
commercial, 12, 19, 20, 23, 27, 28, 41, 56, 83, 105, 108, 109, 116
commercial ties, 28
commodity, 71, 79, 82
common rule, 25, 39
communication, 2, 3, 6, 23, 34, 45, 58, 59, 87, 91, 103
competition, 2, 25, 30, 36, 43, 86, 87, 91
competitive advantage, 27, 41, 106
competitiveness, 1, 3, 87, 99, 104
competitors, 102, 105, 106
complexity, 22, 37, 43, 92, 102, 104
computer, 16, 17, 18, 33, 45, 58, 59, 80, 91, 92, 102, 104, 111, 115
computing, 7, 15, 16, 28, 45, 85, 88, 104, 107, 110
concordance, 64, 71, 72
confidentiality, 116
Congress, vii, 1, 2, 3, 13, 37, 42, 43, 47, 50, 51, 52, 85, 86, 87, 94, 99, 100, 102, 105, 108, 111, 113, 116
connectivity, 41, 45
consensus, 12, 32
Consolidated Appropriations Act, 111
construction, 58, 74, 80, 112, 114
consulting, 59, 75, 80
consumer protection, 36, 37, 42
consumers, 2, 3, 5, 8, 9, 16, 18, 19, 23, 30, 57, 61, 99
consumption, 7, 100, 105
cooperation, 29, 36, 37, 39, 40, 42, 52
coordination, 7, 52
cost, 18, 27, 88, 95, 96, 104
cost of living, 96
Council of the European Union, 25
counterterrorism, 28
covering, 38, 84
CPC, 49
criticism, 21
cross-border data flows, vii, 1, 3, 15, 24, 25, 35, 36, 38, 39, 41, 43, 55, 61
cross-border trade, vii, 54, 55, 57, 58, 59, 62, 64, 67, 71, 72, 73, 74
Cross-Border Trade, v, 53, 60, 67
cryptography, 37
customers, 8, 11, 13, 15, 17, 21, 23, 59, 69, 87, 92
Customs and Border Protection (CBP), 42, 99
cybersecurity, 11, 15, 23, 28, 29, 37, 39, 41, 44, 48
cycles, 88, 94

D

data analysis, 45
data center, 16
data collection, 46
data communication, 87
data processing, 36, 87
Department of Commerce, 2, 8, 11, 12, 17, 43, 45, 47, 53, 55, 70, 114
Department of Defense, 105, 107, 108, 115, 116
Department of Energy, 107
Department of Homeland Security, 50
Department of Justice, 27, 49
developed countries, 4, 33
developing countries, 4, 31, 32, 36, 42
dialogues, 36
digital communication, 59
digital divide, 7, 9
digital economy, vii, 1, 3, 11, 25, 30, 35, 39, 40, 41, 43, 45, 53, 55, 58, 61

Digital Economy, v, 12, 41, 43, 44, 45, 46, 47, 50, 53
digital technologies, 7, 18, 54, 58, 83
Digital Trade, v, vii, 1, 2, 3, 4, 6, 7, 8, 9, 10, 11, 12, 13, 14, 15, 19, 20, 21, 23, 24, 30, 32, 34, 35, 36, 37, 38, 39, 40, 41, 42, 43, 44, 45, 46, 47, 48, 51, 52, 56, 57, 68, 83
digital trade agreement provisions, vii, 3
Digitally-Deliverable Services, v, 53, 58, 60, 63, 64, 66, 68, 74
diodes, 88, 104
direct measure, 54, 56
disclosure, 37
distance learning, 56
distribution, 18, 20, 50
Doha, 30, 31, 34, 35, 50, 51
DOI, 44, 45, 47, 50

E

East Asia, 99, 100
e-commerce, 3, 7, 8, 12, 15, 25, 26, 30, 31, 32, 33, 35, 36, 37, 38, 39, 40, 42, 64, 65, 73, 83
economic development, 16
economic growth, 10, 25, 36, 41, 43
economic integration, 61
economies of scale, 62, 102
electricity, 88, 104
employees, 9, 61, 95, 96
employment, 8, 94, 95, 111
encryption, 20, 23, 34, 49
energy, 103, 107, 108
enforcement, 20, 23, 25, 29, 32, 33, 35, 37, 42, 43, 44
engineering, 55, 59, 75, 80, 85, 86, 96
environment, 13, 16, 18, 25, 32, 33, 34, 40, 42, 53, 55
equipment, 14, 17, 72, 75, 78, 80, 82, 87, 92, 95, 96, 97, 101, 105, 111
espionage, 27, 99
Europe, 6, 54, 67, 84, 86, 89, 93, 95, 97, 98, 101, 103, 114

European Commission, 24, 25, 39, 49, 51, 103, 114
European Parliament, 25, 39, 51
European Union, 21, 22, 23, 24, 25, 52
evidence, 65
exclusion, 72
executive branch, vii, 1, 3
Executive Order, 115
expenditures, 94, 95, 112
export market, 97
exports, 8, 12, 13, 18, 24, 31, 39, 43, 54, 56, 60, 62, 63, 64, 65, 67, 68, 69, 71, 72, 73, 74, 83, 86, 97

F

fabrication, 86, 89, 92, 95, 96, 97, 101, 103, 107, 108, 109, 110
Fabrication, 95, 110
Federal Communications Commission, 23
federal government, 85, 88, 103, 106
Federal Register, 114, 115
financial, vii, 1, 3, 4, 15, 23, 30, 32, 36, 37, 38, 40, 55, 57, 58, 59, 60, 61, 82, 88, 100
flexibility, 15, 36
force, 14, 15, 22, 24, 30, 31, 34, 36, 43, 105
foreign companies, 21, 22, 28, 101, 102
foreign direct investment (FDI), 5, 28
foreign firms, 28
foreign investment, 29
foreign person, 102
formation, 42, 99, 105
foundations, 45
France, 20, 22, 48, 52, 69, 84, 103
free trade, 13, 14, 51
freedom, 11
funding, 7, 44, 85, 87, 88, 99, 100, 103, 104, 106
funds, 59, 60, 79

G

GAO, 111, 116
GDP, 2, 4, 8, 9, 10, 12, 18, 24
General Agreement on Trade in Services (GATS), 30, 33, 35
Germany, 7, 20, 52, 69, 84, 114
Global Competitiveness Report, 46
global demand, 91
global digital trade policy, vii, 1
global economy, 2, 89
global leaders, 87
global markets, 8
global trade, vii, 1, 7, 11, 31
globalization, 2, 6, 44
goods and services, 4, 7, 25, 40, 50, 54, 60, 71, 116
Google, 9, 16, 22, 27, 46, 47, 48, 54, 57, 59
governance, 9, 12, 14, 26
government procurement, 16, 21, 28, 37, 38
governments, 3, 8, 15, 16, 17, 19, 30, 34
grants, 100
gross domestic product, 2, 4, 18, 62
growth, 2, 4, 7, 10, 23, 25, 30, 35, 36, 41, 43, 46, 47, 66, 92, 100, 101, 107, 114
guidance, 26, 83
guidelines, 17, 29, 38

H

harmonization, 25, 29
health, 38, 62, 69, 72, 76, 77, 81, 85, 99
health care, 62, 76, 81
health practitioners, 72, 81
high performance computing (HPC), 107
House, 29, 44, 49, 70, 86, 109, 113, 116
housing, 77, 81
human capital, 9
hybrid, 16, 38
hydrogen atoms, 109

I

imports, 8, 14, 54, 55, 60, 61, 67, 68, 69, 98, 99
India, 7, 17, 20, 21, 31, 38, 52, 68, 69, 84
indigenous innovation, 21, 28
individuals, 5, 8, 14, 43
Indonesia, 16, 52, 84
industrial policies, 3
industrial sectors, 94
industry(ies), 9, 12, 15, 16, 17, 19, 25, 26, 29, 38, 39, 42, 46, 59, 62, 64, 65, 66, 69, 70, 71, 72, 73, 74, 81, 82, 83, 85, 86, 87, 88, 89, 90, 91, 92, 93, 94, 96, 97, 98, 99, 100, 101, 102, 103, 104, 105, 106, 107, 109, 110, 111, 112, 113, 114, 115, 116
Information and Communication Technologies, 37, 51
information economy, 30
information sharing, 29, 32
information technology, 86, 98
infrastructure, 5, 9, 12, 15, 16, 25, 41, 44, 45
initiation, 104
insertion, 116
integrated circuits, 14, 90, 91, 92, 100, 108, 110, 111, 112
integration, 40, 61
integrity, 61
intellectual property, 2, 3, 12, 14, 15, 17, 27, 28, 37, 57, 59, 61, 96, 99
intelligence, 23, 108
intelligence gathering, 23
intermediaries, 7
International Monetary Fund, 58, 83
international standards, 21, 83
international trade, vii, 1, 3, 4, 7, 8, 18, 22, 24, 41, 58, 61, 69
Internet, vii, 1, 2, 3, 4, 5, 7, 8, 9, 11, 12, 15, 16, 18, 19, 20, 22, 26, 27, 30, 31, 33, 34, 36, 37, 39, 41, 42, 44, 45, 46, 48, 49, 50, 51, 53, 54, 55, 57, 59, 60, 61, 62, 69, 70, 74, 80, 83, 84, 87

Index

investment, 9, 20, 25, 28, 41, 79, 88, 94, 96, 103, 107
investments, 13, 79, 82, 85, 104
IPR, 2, 3, 15, 17, 18, 19, 20, 21, 32, 33, 34, 37, 40, 44, 99
Ireland, 7, 55, 69, 84
issues, vii, 2, 3, 12, 14, 19, 20, 23, 25, 26, 27, 31, 32, 33, 34, 35, 37, 39, 40, 41, 42, 43, 104
Italy, 20, 52, 84, 103

J

Japan, 38, 41, 51, 52, 55, 68, 69, 84, 87, 89, 93, 95, 97, 99, 100, 101, 105, 113, 115
joint ventures, 28

K

Korea, 23, 36, 51, 52, 84, 87, 89, 95, 97, 100, 101, 102, 113

L

labor market, 9
Latin America, 6
law enforcement, 23, 29, 35, 43, 44
laws, 2, 3, 14, 20, 25, 26, 28, 37, 38, 39, 57
laws and regulations, 28
LDCs, 33
learning, 45, 56, 83
LED, 101, 112
legal protection, 34, 96
legislation, vii, 1, 23, 37, 99
liberalization, 29, 38, 39
license fee, 58, 59, 60, 62, 68, 78, 82
lithography, 111
localization, 2, 13, 15, 16, 21, 36, 37, 38, 39, 40, 43, 47, 61, 62
low-interest loans, 100

M

machine learning, 45
majority, 39, 54, 90, 92, 99
Malaysia, 84, 93, 97, 112
management, 22, 34, 45, 76, 80, 81
manufacturing, vii, 7, 16, 20, 46, 86, 87, 92, 93, 94, 95, 96, 97, 100, 101, 102, 103, 105, 106, 107, 108, 109, 110, 111, 112, 116
market access, 8, 13, 15, 21, 22, 28, 30, 35, 37, 38, 40, 43
materials, 19, 59, 85, 87, 96, 97, 105, 108
media, 5, 7, 9, 21, 22, 44, 46, 48, 57, 65, 84, 102, 113
medical, 14, 31, 73, 76
memory, 87, 91, 102, 110
methodology, 63, 70, 72
Mexico, 14, 34, 41, 52, 84, 97
microelectronics, 107, 108, 109, 116
military, 56, 85, 87, 88, 105, 108, 116
miniaturization, 107
mobile telecommunication, 103
models, 2, 9, 87
moratorium, 32
multilateralism, 35
multinational companies, 17
multinational firms, 15
music, 18, 19, 56, 57, 59, 65, 66, 87

N

National Academy of Sciences, 96, 106
national borders, 61
National Research Council, 112, 115
national security, vii, 1, 2, 3, 15, 23, 24, 28, 38, 41, 43, 86, 99, 102, 108, 109, 116
National Security Agency, 116
National Security Council, 49
nationality, 28, 61
negative consequences, 27
negotiating, 35, 37, 39, 40

negotiation, 39
Netherlands, 84
network congestion, 23
New Zealand, 47, 84
Nigeria, 16, 17
nontariff barriers, 2, 12, 15, 32, 39
North America, 71, 95, 111
North Korea, 23
Norway, 84

O

Obama Administration, 27
Obama, President Barack, 27, 107
Office of the United States Trade Representative, 47
operations, 20, 22, 29, 89, 96, 103
opportunities, 2, 9, 26, 28, 29, 30, 41, 42
optoelectronics, 90
Organization for Economic Cooperation and Development (OECD), 2, 5, 7, 15, 30, 41, 44, 45, 47, 48, 50, 52, 62, 63
outpatient, 73, 76, 81
oversight, vii, 1, 3, 43, 44

P

Pacific, 2, 3, 14, 35, 36, 38, 42, 47, 51, 52, 54, 67, 84, 86, 98
Panama, 51
Parliament, 25, 39, 51
participants, 14, 38, 98, 106
patents, 17, 32, 59, 98
penalties, 20, 26, 37
performers, 32, 81
physicians, 72, 76, 81
piracy, 18, 19, 34
plants, 87, 95, 96, 101
platform, 4, 8, 25, 61
policy, vii, 1, 2, 3, 9, 11, 15, 16, 20, 22, 30, 36, 38, 41, 42, 43, 50, 53, 55, 67, 86, 102, 109, 115
policy issues, vii, 3

policy options, 109
policymakers, 11, 23, 25, 43, 53, 55, 61
population, 5, 24, 69
portability, 25
potential benefits, 9
President, 26, 27, 46, 49, 51, 107, 112, 115, 116
price changes, 94
price index, 94, 111
principles, 29, 30, 33, 41, 42, 43
private enterprises, 100
private sector, 11, 43, 99
procurement, 16, 21, 28, 37, 38
producers, 12, 32, 87, 88, 92, 94, 96, 97, 98, 100, 102, 112
product coverage, 32
product design, 105
profit, 89, 115
programming, 19, 80, 82
project, 63
property rights, 2, 3, 17, 37, 99
prosperity, 46
protection, 16, 17, 20, 24, 25, 26, 29, 32, 33, 34, 35, 36, 37, 38, 39, 40, 41, 42, 49, 50, 61, 96, 99
protectionism, 11
public figures, 81
public policy, 15, 36, 38, 42, 43, 53, 55
publishing, 22, 48, 64, 80

Q

quality control, 89
quantum computing, 89, 110
quantum phenomena, 109, 110

R

radio, 7, 14, 45, 66, 91
real estate, 75
real wage, 8
reciprocity, 37
recommendations, iv, 32, 109
recreation, 81

recreational, 58, 81
Reform, 45, 49, 112
regulations, 12, 13, 14, 16, 22, 25, 28, 38, 50, 96
regulatory agencies, 42
regulatory requirements, 42
rehabilitation, 81
reinsurance, 59
reliability, 43, 96, 104, 108
remediation, 76, 81
reproduction, 20
requirements, 2, 13, 15, 16, 17, 21, 22, 25, 26, 36, 37, 38, 40, 42, 61, 62, 71, 116
research funding, 103
resources, 45, 59, 86, 106, 112, 116
response, 53, 55
restrictions, 13, 15, 16, 28, 37, 61
restructuring, 110
revenue, 89, 94, 100, 103
rights, 2, 3, 17, 32, 33, 34, 37, 61, 66, 99
rules, vii, 1, 2, 3, 11, 12, 14, 22, 25, 26, 28, 35, 37, 39, 40, 42, 48, 49
Russia, 16, 19, 22, 48, 52

S

safety, 38, 43, 99
Samsung, 91, 96, 100, 101, 106
sanctions, 27
Saudi Arabia, 52
scope, 31, 40, 83
Secretary of Commerce, 46
Secretary of Defense, 108
Secretary of the Treasury, 102
security, vii, 1, 2, 3, 12, 15, 16, 23, 24, 28, 29, 35, 38, 41, 43, 61, 62, 81, 86, 99, 102, 108, 109, 116
security services, 81
semiconductor, vii, 85, 86, 87, 88, 89, 90, 91, 92, 93, 94, 95, 96, 97, 98, 99, 100, 101, 102, 103, 104, 105, 106, 107, 108, 109, 110, 111, 112, 113, 114, 115, 116

semiconductors, 13, 31, 85, 86, 87, 88, 89, 91, 92, 94, 97, 98, 99, 100, 101, 102, 103, 104, 107, 108, 109, 110, 111, 112, 113, 115
servers, 16, 17, 19, 45, 62
service industries, 64, 65, 71
service provider, 15, 16, 20, 34, 37, 45, 74, 80
services, 1, 3, 4, 6, 7, 8, 9, 11, 12, 13, 14, 15, 16, 17, 18, 19, 20, 22, 23, 24, 25, 27, 28, 30, 35, 36, 37, 38, 39, 40, 42, 45, 46, 48, 50, 54, 55, 56, 57, 58, 59, 60, 61, 62, 63, 64, 65, 66, 67, 68, 69, 70, 71, 72, 73, 74, 75, 76, 77, 80, 81, 82, 83, 93, 95, 109, 114, 116
silicon, 87, 88, 92, 96, 104, 105, 110, 111
Silicon Valley, 115
Singapore, 13, 36, 84
small business, 11, 62
society, 17, 21, 25, 42
software, 16, 18, 19, 45, 59, 87, 91, 115
solution, 9, 104
South Korea, 36, 51, 52, 87, 89, 91, 95, 97, 100, 101, 102, 113
Southeast Asia, 97, 101
sovereignty, 22, 26
spending, 6, 66, 88
spillover effects, 22
state, 2, 11, 25, 27, 45, 46, 95, 102, 116
statistics, 49, 54, 55, 56, 57, 58, 70, 73, 83, 94, 100, 101, 111, 113
suppliers, 12, 13, 14, 27, 28, 86, 100, 108, 109, 111, 116
supply chain, 7, 9, 54, 56, 62, 70, 71, 86, 98, 101, 107, 109
support services, 76, 81
surplus, 8, 18, 54, 60, 61
surveillance, 23, 24
Sustainable Development, 44, 45
Switzerland, 19, 55, 67, 84

T

Taiwan, 84, 87, 89, 95, 97, 100, 101, 102, 113
tariff, 2, 12, 14, 22, 37, 39, 47, 86, 98, 112
Task Force, 53, 55, 70, 105, 115, 116
technology(ies), 3, 6, 7, 8, 9, 10,12, 17, 18, 20, 21, 22, 28, 32, 33, 35, 37, 39, 42, 54, 55, 56, 58, 62, 83, 85, 86, 87, 88, 89, 91, 97, 98, 101, 102, 104, 105, 106, 107, 108, 109, 110, 112, 113, 114, 115
technology transfer, 33, 35, 37
telecommunications, 28, 30, 37, 38, 44, 46, 60, 61, 69, 79, 80, 97, 103
testing, 2, 15, 22, 92, 98, 109
Thailand, 84, 113
theft, 2, 18, 19, 22, 23, 27, 29, 40, 41, 99
TPA, 13, 37, 47
trade, vii, 1, 2, 3, 4, 5, 6, 7, 8, 10, 11, 12, 13, 14, 15, 17, 18, 19, 20, 21, 22, 23, 24, 25, 26, 27, 29, 30, 31, 32, 34, 35, 36, 37, 38, 39, 40, 41, 42, 43, 44, 45, 46, 47, 49, 50, 51, 54, 55, 56, 57, 58, 59, 60, 61, 62, 63, 66, 67, 68, 69, 71, 72, 73, 74, 83, 84, 86, 95, 96, 97, 98, 99, 112, 113
trade agreement, vii, 1, 3, 13, 14, 15, 29, 30, 35, 41, 51
trade negotiations, vii, 1, 3, 13, 30, 32, 36, 41, 44
trade policy, vii, 1, 2, 3, 30, 43, 67
trade rules, vii, 1, 2, 39
trading partners, 34, 41, 69
training, 21, 56, 83, 103
transaction costs, 74
transactions, 3, 15, 54, 55, 56, 57, 61, 74, 83, 102
transmission, 20, 32, 59, 89
transparency, 14, 30, 36, 40
transport, 108
transportation, 109
treaties, 32, 34
treatment, 13, 33, 35, 38, 76

U

U.S. Bureau of Labor Statistics, 111
U.S. Department of Commerce, 47, 53, 70
U.S. economy, vii, 3, 9, 27, 42, 43, 69, 93
U.S. Semiconductor Manufacturing, v, 85, 93
United Kingdom, 49, 52, 55, 68, 69, 84
United Nations, 14, 49, 58, 83
United States, 2, 5, 6, 7, 8, 10, 11, 12, 13, 14, 16, 23, 24, 25, 26, 27, 28, 29, 30, 31, 34, 35, 36, 37, 38, 39, 40, 41, 42, 44, 46, 47, 49, 51, 52, 54, 55, 56, 57, 60, 61, 67, 68, 69, 70, 74, 83, 84, 85, 86, 87, 89, 92, 94, 95, 96, 97, 101, 102, 107, 108, 112, 115

V

vehicles, 79, 82, 87
Venezuela, 84
Vietnam, 14, 16, 20, 31, 47

W

word processing, 87, 91
workers, 8, 9, 10, 12, 62, 86, 94
workforce, 41, 94
World Bank, 4, 5, 7, 9, 44, 45, 46, 49
World Development Report, 5, 44, 45,
World Trade Organization (WTO), 2, 13, 14, 28, 30, 31, 32, 33, 34, 35, 36, 37, 38, 39, 43, 47, 50, 51, 62, 63, 86, 98, 112, 113
worldwide, 4, 6, 12, 22, 26, 42, 86, 87, 90, 92, 95, 96, 97, 99, 100, 101, 103